# The Transformation and Turnaround of Employers' Federation

# SEIFSA

Kaizer M. Nyatsumba

UJ Press

*The Transformation and Turnaround of Employers' Federation SEIFSA*

Published by UJ Press
University of Johannesburg
Library
Auckland Park Kingsway Campus
PO Box 524
Auckland Park
2006
https://ujpress.uj.ac.za/

https://doi.org/10.36615/9780906785249
978-0-906785-37-9 (Paperback)
978-0-906785-38-6 (PDF)
978-0-906785-39-3 (EPUB)
978-0-906785-40-9 (XML)

This publication had been submitted to a rigorous double-blind peer-review process prior to publication and all recommendations by the reviewers were considered and implemented before publication.

Language Editor: Richard Bowker
Cover design: Hester Roets, UJ Graphic Design Studio
Typeset in 9/13pt Merriweather Light

# Contents

# Foreword

When Dr Kaizer Nyatsumba shared the manuscript of this book with me and asked whether I could write this foreword, I was excited. I was pleased to be given this opportunity to share some of my experiences, first, at the Industrial Development Corporation, where I was Chief Economist for several years, and, second, at the University of the Witwatersrand over the past few years.

This book is ground-breaking (it is a pioneer, "*imvulandlela*", in the Nguni languages of South Africa). Nyatsumba presents a historiography of South Africa's industrialisation with a scope that ranges from the organisational to the individual. In doing so, he makes a significant contribution to our understanding of the process from a black perspective, that is, from the point of view of those who were historically excluded from the mainstream and the commanding heights of the economy. He is the first black Executive Director of the Steel and Engineering Industries Federation of Southern Africa (SEIFSA) – and so far, the only one – to hold that position in the organisation's 80-year history. It is in this light that I share what I have learnt from this significant historiography by Nyatsumba.

In this book, Nyatsumba tells the story of an organisation rooted in South Africa's economic infrastructure and manufacturing-capability creation, of its successes in relation to the mineral and energy complex during apartheid and its decline under democratic rule as a result of conflicting interests – and of the turnaround of the federation that has long represented employers in the metals and engineering (M&E) sub-sector. Nyatsumba complements this historiography by providing an account of the significance of the metals and engineering manufacturing sub-sector from the perspective of this massive organisation, SEIFSA.

In the introduction, Nyatsumba locates South Africa's industrialisation in relation to the structural transformation of the manufacturing sector:

> In South Africa, manufacturing did not take off until
> the mining industry – which, together with agriculture,
> was the mainstay of the local economy until the 1920s
> – increased demand for manufactured products such as
> processed foods and textiles.

In this passage, Nyatsumba borrows from Sir Arthur Lewis and Nicholas Kaldor, pioneers of the classical economic development perspective, which emphasises the transformation of production and employment as a driving force for growth. In this view, development involves a shift in the structural composition of the economy; in this case, it is away from agriculture and mining to value addition. The industrialisation of South Africa was underpinned by the apartheid state's ambition and its commitment to building new industries through protection and self-sufficiency. In 1922, the state introduced the Eskom Act, which led to transfer of many private-sector power-generation assets to Eskom in 1948 and hence to its dominance of the electricity value-chain. Arguably, this dominance was informed by Eskom's mandate to provide electricity at the lowest possible cost, access to cheap steel from the state-owned Iron and Steel Corporation (Iscor), and the abundance of cheap black labour. The combination of cheap electricity, steel, and black labour underpinned South Africa's structural transformation. Arguably, the subsequent rising costs associated with these factors may account for the South African economy's decline.

According to Fortunato (2022), during the apartheid era in South Africa, the government championed manufacturing by providing various incentives and by protecting against imports through the imposition of economic tariffs. From 1960 to 1980, during which the state drove industrialisation, the manufacturing sector accounted for 30% of South Africa's gross domestic product (GDP) and was by far the biggest employer. During that period, more than half (55%) of the economically active population was employed in manufacturing, with retail the second-largest employer, at 40%. As a percentage of GDP, manufacturing value added averaged 22% (Fortunato, 2022) during that period. By the end of 2022, the manufacturing sector's contribution to GDP was approximately 12.27%, with the significant decline in the sector attributed to South Africa's premature deindustrialisation.

In the past ten years, there has been a resurgence of political-economy research into the premature deindustrialisation of the South African economy. Gilad Isaacs (2018) provides an account of the shift by South African manufacturing firms from production to finance, which he refers to as financialisation. Nimrod Zalk (2017) also picks up the financialisation baton in his account of how many productive firms became financial firms, leading to a decline in manufacturing. In South Africa's manufacturing history, the state has played a pivotal role, as argued by Clarke (1994); in relation to the minerals and energy complex by Fine and Rustomjee (2018); in Ballim's (2023) work on power, politics, and the apartheid state through the lens of Eskom; and in Mondi's (2021) historical account of the restructuring of Eskom. Underpinning this work is the significant contribution to industrialisation of the state-owned development finance institution, the Industrial Development Corporation.

Nyatsumba investigates manufacturing's trajectory through the lens of SEIFSA and attributes its emergence and development to General Jan Smuts:

> The Smuts government sourced most of its war-time requirements from South Africa. According to SEIFSA (n.d.), during the war years (from 1940 to 1944), South African M&E companies manufactured and delivered:
>
> - More than 57 000 armoured fighting vehicles and 35 000 military motor vehicles;
>
> - Over 70 000 tonnes of bombs, 50 000 tonnes of shells and shell casings, and 28 000 tonnes of mortar bombs, grenades and landmines;
>
> - Nearly 20 000 tonnes of small arms and ammunition; and
>
> - Over 1.5 million helmets.

Nyatsumba's view is supported by SEIFSA's articulation of the sub-sector's history:

> Out of these wartime facilities came aircraft hangars, bridges, travelling cranes, barges, rock drills, pumps and valves, electric motors, transformers, field hospital equipment, ships' stores for the British Admiralty, thousands of kilometres of telephones and telegraph wire for the signal services ... the list is almost endless.

> The Western Desert pipeline supplying water for allied troops in North Africa was built largely from South African metals.

The M&E sub-sector had transitioned to manufacturing products that had previously been imported during South Africa's import-substitution industrialisation. For this period, Nyatsumba provides a picture of a thriving and peaceful industry focused on profits and, according to SEIFSA, the welfare of its employees. In the 1950s, Nyatsumba explains, relations between employers, as represented by SEIFSA, and the labour unions were "amicable and constructive", in SEIFSA's words, thanks to the existence of the National Industrial Council. One of the products of these "amicable and constructive" relations between employers and labour was the formation, in the mid-1950s, of the industry's Sick Pay Fund and the Group Life and Provident Fund for skilled workers.

The National Party's victory in 1948 over Smuts's South African Party led to significant social, political, and economic changes. In 1953, the National Party government passed the Black Labour Relations Regulation Act, which excluded black people from participation in any form of collective bargaining. From that time on, black workers' interests were to be dealt with through a white-controlled central board. In addition to providing for centralised collective bargaining between employers and unions representing white workers, the 1924 Industrial Conciliation Act also entrenched discrimination at the workplace.

In the late 1960s, the Black Consciousness Movement led by Stephen Bantu Biko emerged. Although the movement's plans to launch unions were unsuccessful because of the repressive apartheid regime, it was, however, successful in building community-based organisations, and it rejected joint black–white political activity. This pushed white intellectuals and aspirant activists to attend to union issues. In January 1973, workers in Durban, with no obvious help from unions or any other body, went on strike. The spontaneity of this event points to structural conditions that allowed for strikes to occur in many factories and workplaces in the city. The wave of strikes proved to be a puzzle for many: the government, the media, white unions, and social scientists, among others.

Members of SEIFSA struggled to fill vacancies because of skills shortages, despite the federation's involvement in the recruitment and screening of white immigrants seeking to be employed in the M&E sub-sector. In 1972, SEIFSA reached an agreement with its counterpart trade unions to allow black workers to occupy some jobs previously reserved for whites alone. In 1973, the Metals and Allied Workers Union was established with the assistance of the General Factory Workers Benefit Fund. The 1976 students uprising in Soweto and the subsequent work stay-aways saw that dispensation extended to include positions just below those of artisans.

Faced with industrial instability and community struggles, in 1977, the apartheid regime set up the Wiehahn Commission. Among the commission's key recommendations were that full freedom of association to be extended to all employees, that trade unions be allowed to be registered, regardless of their racial composition, and that statutory job reservation be phased out. On 1 December 1985, the Congress of South African Trade Unions (COSATU) was formed under the slogan of "one industry, one union". The organisation contributed significantly to the South African political transition that led to a democratic dispensation in 1994.

The African National Congress (ANC) inherited a highly indebted government and a fragile, inward-looking economy requiring structural economic reforms. The ANC government adopted a market-friendly macro-economic strategy, the

Growth, Employment and Redistribution Strategy (GEAR), to the irritation of unions and many on the left who had contributed to the struggle against apartheid. However, GEAR led to significant inward investment, employment, and the longest sustained economic growth cycle in South Africa, averaging 4.1% during President Thabo Mbeki's term in office (South African Market Insights, 2018).

Unfortunately for Nyatsumba, he took the higher office at SEISFA in 2013 in a period of populist rhetoric and State capture under President Jacob Zuma. Under the chairmanship of Ufikile Khumalo, who led an IDC subsidiary affiliated to SEIFSA, Nyatsumba succeeded in transforming the organisational structure of SEIFSA, but could not turn the fortune of its members around. As Nyatsumba observes, just as the apartheid state had a significant role in the industrialisation and structural transformation of South Africa, the democratic government also played a role in the weakening of the country's industrial and structural transformation. The ANC Government failed to complete the restructuring of Eskom, which had been proposed in the 1998 Energy Policy Paper, and took away a number of prospecting rights licences from unempowered companies in terms of the new mining law of 2003, the Mineral Resources and Petroleum Development Act, thereby disrupting the minerals and energy complex. Following power blackouts in 2007, the South African government brought the state back into the economy, and it subsequently became the centre of corruption.

Nyatsumba worked on rebuilding State-SEIFSA relations:

> Times were tough and, for many in manufacturing, it felt like they had been deserted by the Government. Did Pretoria care about their feelings? If so, how did it respond?

Unfortunately, it was just sweet talk, with no actions occurring on the ground. If state–SEIFSA relations were bad, labour–SEIFSA relations worsened after the Association of Mineworkers and Construction Union's (AMCU's) five-month-long strike in the platinum-mining sector at the beginning of 2013. The strike set the tone for the approach that National Union of Metalworkers of South Africa (NUMSA) and other unions adopted in the M&E negotiations of the same year. Seen as the most radical union until then, NUMSA was worried about the effect that the more radical AMCU would have on its membership in other economic sectors to which it might extend its reach. Even then, there were fears that AMCU was eyeing the M&E sub-sector with a view to joining the Metal and Engineering Industries Bargaining Council as a party to negotiations.

The Steel and Engineering Industries Federation of Southern Africa and its members found themselves between a rock and hard place as relations with labour and the Government deteriorated. These tensions also filtered down into SEIFSA, and Nyatsumba's position in the organisation deteriorated, leading to his ousting by a black-led board of SEIFSA. Although the changing of chairs was associated with the state-owned entities and government departments, it had also diffused to private-sector organisations.

The real tale of Nyatsumba's book is of the deterioration in South Africa. The book is a tale of how the collapse of national infrastructure, deteriorating governance, a lack of accountability, and corruption have become institutionalised. Nyatsumba shares his experiences and – reading from his experience – one can come to a conclusion that the progressive structural transformation of the South African economy is now a thing of the past. With the institutionalisation of Black Economic Empowerment and the appetite of black industrialists to import M&E services – as indicated in the works of Zalk (2017), Isaacs (2018) and Seeraj Mohamed (2019) on the externalisation of productive firms and their becoming financially focused, and in Nzeni Netshintomboni's (2021) work on the significant role of the IDC in the industrialisation process – has manufacturing withered?

Nyatsumba provides an optimistic perspective – that there is light at the end of the tunnel if we can address our challenges and learn from the past – as articulated by Ballim (2023) in relation to the significant contribution of Iscor and Eskom in the Waterberg, the challenges associated with the new Medupi power station, and Mondi's (2021) optimism regarding the role of a just transition in revitalising manufacturing.

**Dr Lumkile Mondi**
Senior Lecturer: Wits Business School,
University of the Witwatersrand

## References

Ballim, F. (2023). *Apartheid's leviathan: Electricity and the power of technological ambivalence*. Athens, OH: Ohio University Press. https://doi.org/10.2307/jj.9669730

Clark, N. L. (1994). *Manufacturing apartheid: State corporations in South Africa*. New Haven, CT: Yale University Press. https://doi. org/10.2307/j.ctt2250wrc

Fine, B. & Rustomjee, Z. (2018). *The political economy of South Africa: From minerals–energy complex to industrialisation*. London: Routledge. https://doi.org/10.4324/9780429496004

Fortunato, A. (2022). Getting Back on the Curve: South Africa's Manufacturing Challenge. The Growth Lab at Harvard University. Center for International Development Research Fellow & Graduate Student Working Paper 139. https://doi.org/10.2139/ssrn.4411854

Isaacs, G. L. (2018). Financialisation in Post-Apartheid South Africa. PhD thesis. London, SOAS University of London.

Kaldor, N. (1957). A model of economic growth. *The Economic Journal*, 67(268):591–624. https://doi.org/10.2307/2227704

Lewis, W. A. (1954). Economic development with unlimited supplies of labour. *The Manchester School of Economic and Social Studies*, 22(2):139–191. https://doi.org/10.1111/j.1467-9957.1954.tb00021.x

Mondi L. (2021). A Long Walk to a Just Energy Transition – The Political Economy of the Restructuring of Eskom: 1985 and Beyond. PhD thesis. Johannesburg: University of the Witwatersrand.

Mohamed, S. (2019). The Political Economy of Accumulation in South Africa: Resource Extraction, Financialization, and Capital Flight as Barriers to Investment and Employment Growth. PhD thesis. Amherst: University of Massachusetts.

Netshitomboni, N. (2021). The Industrial Development Corporation under Apartheid: Financing the Industrialisation of South Africa, 1940 to 1990. PhD thesis. Johannesburg: University of the Witwatersrand.

South African Market Insights. (2018). *South Africa's economic performance under Thabo Mbeki*. https://www.southafricanmi.com/thabo-mbeki-sa-economy-6sep2018.html

Zalk, N. (2017). The Things We Lost in the Fire: The Political Economy of Post-Apartheid Restructuring of the South African Steel and Engineering Sectors. PhD thesis. London: SOAS, University of London.

# Introduction

Although the main focus of this book is the Steel and Engineering Industries Federation of Southern Africa (SEIFSA), the umbrella organisation representing employers in the metals and engineering (M&E) sub-sector, the economic sub-sector in which that organisation operates is part of the larger manufacturing sector. Of necessity, therefore, the book begins with a chapter on the fortunes of manufacturing in South Africa, followed by another one on the M&E sub-sector – sometimes called the steel and engineering sub-sector – which is itself widely diversified. Only then does it meander its way to the history of SEIFSA, which celebrated it's 80th anniversary in 2023.

The members of SEIFSA are independent employer associations representing different sub-industries or interests in the M&E sub-sector. Individual companies are members of the employer associations, which collectively form the federation. Over the years, the number of affiliated associations has waxed and waned, as have the total number of companies represented by the federation through the associations, and the overall number of employees working for companies represented by SEIFSA.

Various factors have been responsible for this situation, but by far the main one has been the precipitous decline of manufacturing in South Africa, especially since the global financial crisis of 2008/9. This decline has been described by some scholars as "premature deindustrialisation" (Fortunato, 2022). Within the manufacturing sector, the M&E sub-sector suffered the worst fate.

The book touches on the history of SEIFSA, as captured by the organisation itself, before and after the dawn of South Arica's democracy in 1994. However, greater focus is paid to the period from November 2013 to July 2021, when the federation – which had been experiencing growing financial losses until November 2013 – underwent a transformation and had its fortunes turned around following my appointment as its first – and so far only – black Chief Executive Officer (CEO) in its 80-year history.

While drawing from published material on the M&E sub-sector and on SEIFSA, the book also relies on my observations and recollections from my almost eight-year tenure as SEIFSA CEO. During this time, I was rigorous in keeping notes in my personal diaries and notebooks. However, the information included in this book is generally available to scrupulous scholars, students, and observers of the M&E sub-sector, and some of the material has previously been used in my opinion pieces in the mainstream business media; hence, no confidences are being violated. Much of what is captured in this book was reported on in the media or reflected in SEIFSA's own internal publications, such as websites and the periodical, *SEIFSA News*, during that eight-year period.

**Kaizer M Nyatsumba**
October 2024

# Part 1

# Chapter 1

# Manufacturing in South Africa

## Introduction

While its essence has remained the same over the centuries, manufacturing – defined by the *Concise Oxford English Dictionary* variously as the making, production, invention, or fabrication of something that did not previously exist – has taken different forms over the years. With its origin in the French and Italian phrases *manu factum* and *manifattura*, respectively, literally meaning "made by hands", manufacturing took a giant leap forward during the Industrial Revolution when various British industries transformed from small-scale, hand-operated production to sophisticated processes that benefitted from new technologies, such as steam power, and the availability of better infrastructure, such as canals and railways (BakerBaynes, 2017).

Before the Industrial Revolution, which emerged in Britain during the 18th century and spread across the globe, what passed for manufacturing was mostly intended for home use or for use within a village, and production took place in small workshops that employed few people who used hand tools and craft methods (Wilson, 2000). In the era of the fourth industrial revolution – defined by World Economic Forum founder Klaus Schwab (2016) as "the fusion of technologies that is blurring the lines between the physical, digital and biological spheres" – manufacturing has become more sophisticated, benefitting from ever more advanced technologies and economies of scale and scope, which have considerably reduced the unit cost of production.

In South Africa, manufacturing did not take off until the mining industry – which, together with agriculture, was the mainstay of the local economy until the 1920s – increased the demand for manufactured products such as processed foods and textiles. From its inception, South African manufacturing enjoyed considerable support from the state. To stimulate manufacturing, the government of the day set up state-owned entities in 1922 and 1928 that provided cheap electricity and steel, respectively, and imposed import tariffs to protect the domestic sector from dumping by Western powers after the First World War. In addition, other state-owned companies were set up to produce paper, fertiliser, textiles, chemicals, oil and arms in order to promote self-sufficiency (BakerBaynes, 2017; Coutsoukis, 2004).

So sophisticated has manufacturing now become that the United States National Science Foundation (n.d.) describes it as "the process of converting dreams into objects that enrich lives"; while the Corporate Finance Institute (2022) defines it as "a system in which raw materials are transformed into finished products while adding value in the process".

## A Brief History: From Colonial Rule to the Introduction of Democracy in 1994

As was typical of colonies at the time, in its early years, the Union of South Africa, as a British colony, was heavily dependent on British imports for manufactured goods. With the local economy heavily dependent on agriculture and gold mining, the little manufacturing that took place locally – such as wine-making, tanning, and tallow production – was based on the beneficiation of agricultural products. It was not until Britain found itself mired in the First World War and unable to produce enough for its colonies that South Africa was forced by circumstances to become self-reliant in order to cater for its own needs. From this starting point, South African manufacturing at a viable scale was actively supported and incentivised by the government (Coutsoukis, 2004; Callinicos, 1987; Fine & Rustomjee, 2018; Wilson, 2000).

In addition to agriculture-based manufacturing, the bustling mining industry, which had come into existence following the discovery of gold on the Witwatersrand in 1886, led to the development of other local manufacturing sub-industries to cater for its demands for products such as processed foods, clothing, and footwear. To meet the growing demand for manufactured products, savvy industrialists and some of the mining firms set up companies to produce what they needed, such as dynamite and cement (Callinicos, 1987).

However, the end of the First World War left Britain in a position where it was again able to take advantage of its industrialisation to focus on manufacturing on a massive scale to supply colonies like South Africa. Disentangled from the war, Britain and countries like the USA wanted markets for their products, which often landed in South Africa at a cheaper price than local manufacturers could offer. This dumping led to a situation in which, by the 1920s, South Africa was again more heavily dependent on imports than on local manufacturers (Callinicos, 2000; Coutsoukis, 2004; Fine & Rustomjee, 2018; Wilson, 2020).

Starting in 1919, the government of General Jan Smuts actively promoted local manufacturing. First, it recruited Dr Hendrik van der Bijl and had him return to South Africa from the USA to stimulate manufacturing in the country. Consequently, state-owned corporations like Eskom and the Iron and Steel Corporation (Iscor) were established in 1922 and 1928, respectively, to provide cheap electricity and steel to local manufacturers (Jacobs, 1948).

Following the defeat of Smuts in the 1924 elections, subsequent governments continued to prioritise the development of local industries. Through the Tariff Act of 1925, the new government imposed tariffs on imports in an effort to protect local manufacturers from international competition. This protection led to phenomenal growth in the local manufacturing sector to the extent that, by 1931, it accounted for around 17% of the country's growth domestic product (GDP; Callinicos, 1987). Among the key beneficiaries were local farmers, whose products were beneficiated to produce food, canning, footwear, and cigarettes. Profits were often re-invested to develop the country's manufacturing capability (Callinicos, 1987).

Local manufacturing was not unaffected by the global economic slowdown triggered by the decade-long Great Depression, which began in the USA in 1929. During that period, some factories were forced to shut down, notwithstanding the government's "South Africa First" policy. Massive retrenchments took place, with black workers – who were not covered by the Industrial Conciliation Act of 1924, which allowed white workers to form trade unions which would represent them in bargaining with employers – at the front of the queue of those who lost their jobs. However, thanks to its gold production, South Africa was among the first countries to recover from the depression (Swanepoel & Fliers, 2021).

To protect local manufacturers and farmers during the Depression, the government raised import tariffs and provided bonuses to firms and farmers that exported their products, except in the case of gold, diamonds, and sugar (Callinicos, 1987).

To protect white employees in 1928, the government had passed the Wage Act, which required employers to pay "civilised wages" to "civilised [white] workers". During the 1930s, the Wage Act had unintended consequences. Although it partly achieved its goal, that piece of legislation was to spur South African manufacturers to embrace technology by modernising their factories and investing in machinery for the purpose of mass production. Ironically, that "civilised labour" policy had the effect of helping the country to develop its manufacturing industry (Callinicos, 1987) through the use of technology.

Profits generated from increased gold sales underpinned economic growth as the state raised more taxes. The country was able to pay in gold to finance the acquisition of the heavy machinery needed by Iscor. In the 1930s, more mining companies invested directly in the manufacture of engineering equipment, fertiliser, glass, cement, and in the fishing, and pulp and paper industries. To avoid paying the high protective duties and to be able to take advantage of the cheap electricity and steel, some British and American manufacturers opened factories in South Africa. Among these were General Motors, Ford, Nestlé, Cadbury, McKinnon Chain, Dorman Long, Wilcox, Babcock, Dunlop, Firestone, Siemens, Stewarts and Lloyds, Davy Ashmore, and General Electric (Callinicos, 2000).

Callinicos (1987) reports that, by the end of the 1930s, manufacturing had grown at a much faster pace than any other economic sector in the country; by 1940, it was employing 236 000 workers, the majority of whom were black. Like the First World War before it, the Second World War – which broke out in 1939 – was a boon for South African manufacturing. With manufactured imports again limited, the local manufacturing sector increased its level of self-sufficiency. Iscor's production of the iron and steel required for machinery was an advantage, as was the availability of coal and cheap electricity. The boom in gold sales ensured that the country had the capital that it required to finance its growth in manufacturing. Consequently, during the period of the Second World War (1939–1945), South Africa's manufacturing output nearly doubled in value, from R164 million to R316 million, and, by 1943, the sector "was producing more of the country's wealth than gold mining" (Callinicos, 1987).

The manufacturing sector was able to achieve this feat because of massive support from the government and as a consequence of support from the other two main industries, agriculture and mining. In addition, local manufacturing had the two world wars to thank for its phenomenal growth.

During the decade (1936–1946) which included the Second World War, local manufacturing grew by a minimum of 6% per annum. That growth accelerated to an average of 13.3% in the early 1950s, following the National Party government's tightening of import controls after its ascension to power in 1948. From then onwards, most growth in the manufacturing sector – which went on to be quite diverse – occurred in heavy industry, with the metals and engineering (M&E) sector taking the lead (Coutsoukis, 2004).

As manufacturing became more dominant, it also became more capital intensive, notwithstanding the plentiful availability of labour in the country. Such capitalisation was encouraged by the government, either directly through tax incentives or indirectly through the country's state-owned corporations. Capital investments continued during the 1970s, fuelled by investments by state-owned entities (SOEs) like Eskom and Sasol, and more manufacturers adopted technology to reduce production costs so that they would be competitive in the face of imports and to avoid taking on more labour, which was becoming increasingly politicised (Coutsoukis, 2004).

The pre-democracy years presented serious challenges for local manufacturers. Not only were the levels of political activism on the increase in the country, whose townships were increasingly engulfed by violence, but apartheid South Africa also found itself in the eye of the storm as calls for punitive economic sanctions against Pretoria reverberated around the world. The inflation rate shot up and Pretoria found itself unable to service its foreign loans. Where manufacturing output grew by about 3% per annum in 1981, by 1991 it had declined to about 2.5% (Coutsoukis, 2004).

From March 1989, the South African economy experienced a recession, which worsened considerably in 1992 and then eased off in the first half of 1993 (South African Reserve Bank, 1993). In the decade of the dawn of South Africa's democracy, manufacturing in the country fared slightly better than it had in the previous decade, but its value still remained lower than in both industrialised and developing countries during the 1990s (Kaplan, 2004).

## South Africa's Premature De-Industrialisation

As a sector, manufacturing has long been considered to have "special properties" which are unique to it, and which make it particularly important in a country's development. These characteristics have to do with its contribution to economic growth and its ability to create jobs, especially for the low-skilled. Given these "special properties", it follows that poor performance by the manufacturing sector has a negative impact on a country's rate of growth (Bhorat & Christopher, 2017; Fortunato, 2022; Tregenna, 2008).

In policy documents such as the Industrial Policy Action Plan 2011/12–2013/14, which was subsequently reviewed and updated, the South African government has acknowledged manufacturing's potential to create jobs, given its labour-intensive nature (National Planning Commission, 2011).

In addition to the special characteristic of job creation, manufacturing stimulates economic growth by serving as an important source of demand for the services sector in a country. Manufacturing has been found to have a far greater impact on a country's services sector than the services sector has on manufacturing (Bhorat & Christopher, 2017; Fortunato 2022; Tregenna, 2008). That means that good performance by the manufacturing sector has a more positive impact on the services sector than is conversely the case and that a negative performance by manufacturing often has a ripple effect throughout the economy, starting with the services sector.

This is because manufacturing has "strong backward linkages" to services, which means that various services are required, as inputs, to manufacture a product. Capital and labour are but two of these services, among many others. That makes manufacturing an important source of demand for the services sector. However, it also means that the cost and the quality of those service inputs, which constitute intermediate inputs into the manufacturing process, are crucial for the productivity and the competitiveness of the manufacturing sector (Tregenna, 2008).

In South Africa, manufacturing had its heyday during the apartheid era, when it was championed by the government through the provision of various incentives and protection against imports through the imposition of economic tariffs. From 1960 to 1980, when the state drove industrialisation, the manufacturing sector accounted for 30% of South Africa's GDP and was by far the biggest employer. During that period, more than half (55%) of the economically active population was employed in manufacturing, with retail as the second-largest employer, at 40%. Employment in agriculture nosedived during that period, and manufacturing value added as a percentage of GDP averaged 22%. Consequently, the high level of manufacturing employment compensated for the decrease in agriculture employment (Fortunato, 2022).

The composition of manufacturing during those two decades changed considerably. Most of the manufacturing occurred in the M&E and petrochemical sub-industries, which enjoyed generous government incentives and had strong linkages with the mining sector (Fine & Rustomjee, 2018). Until the 1990s, manufacturing in the country was highly capital intensive. That period saw high levels of investment made in capital stock with the result that employment levels grew at a slower pace (Fortunato, 2022).

According to Fortunato (2022), there is a general trend that the total number of people employed in the manufacturing sector, as well as the amount of manufacturing value added, declines as a country grows richer. This leads to a phenomenon known as de-industrialisation. Notably, developing countries have been reaching their peaks of manufacturing employment and manufacturing valued added at lower levels of income than did their developed counterparts (Fortunato, 2022; Tregenna, 2008).

Tregenna (2008) posits that South Africa moved from being a minerals and resource-based economy to being a capital-intensive manufacturing economy without transitioning through a stage of "labour-intensive light industry"; and that in the late 2000s, the country again "leapfrogged" to being a services-oriented economy without "ever having industrialised fully or having 'derived full benefits' from" industrialisation, which would have included higher economic growth and employment levels.

Fortunato (2022) argues that de-industrialisation takes place at a faster pace for newly industrialised countries. This, he argues, is why developing countries "reach their peaks of manufacturing employment and value added at lower levels of income". In South Africa's case, 1981 was the year when the highest number of people were employed in the manufacturing and value-added sector. According to Fortunato (2022), by 1994 the number of people employed in the sector was what could be expected of an economy that was the size of South Africa's at the time, although the level of manufacturing value added was "considerably above the [international] trend" for developing countries until 2008.

Even though overall employment levels in the sector kept increasing until 2008, manufacturing's share of jobs in the economy declined. This was notwithstanding the fact that manufacturing value added continued to be higher than the global trend during this period.

After the founding democratic elections of 1994, some of the country's manufacturers – especially those in the non-ferrous metal, furniture, and footwear industries – recorded impressive levels of capacity utilisation: up to 90% of their factories (Coutsoukis, 2004). However, upon coming to office, the new government threw manufacturing to the wolves. As it moved to integrate the South African economy into the global economy, the Government, as a member of the World Trade Organization (WTO) – removed the tariffs which had long protected local manufacturers. That left local manufacturers unable to compete with their counterparts in China, Vietnam, and Bangladesh (Bhorat & Rooney, 2017; Rodrick 2006).

Bhorat and Rooney (2017) argue that the local manufacturing sector's "lack of dynamism" is cause for concern since "no country has transitioned from middle- to high-income status without the presence of a vibrant manufacturing sector". Manufacturing's decline coincided with the decline of the country's mining sector, which moved from accounting for 12% of the country's GDP in 2001 to representing 8% in 2017. During the same period, manufacturing moved from being the third-largest employment sector, employing 14.7% of the country's working population, to employing on 11.3% of the working population (Bhorat & Rooney, 2017).

From 2008 to 2014, the manufacturing sector lost 331 000 jobs, the highest of any sector in that period. That saw the M&E industries emerging as the largest sub-sector of manufacturing, thanks to projects undertaken in the run-up to the 2010 FIFA World Cup in South Africa (Bhorat & Rooney, 2017).

However, the growth in the manufacturing sector between 1994 and 2008 was accompanied by an increase in investment. During that period, gross fixed-capital formation (GFCF) in South Africa increased by around 15% per annum; however, it slowed down considerably to about 4% thereafter. The M&E sub-sector was the biggest beneficiary of that growth in investment, largely as a consequence of investment in infrastructure. A breakdown in network industries like logistics and electricity supply since the late 1990s, coupled with policy uncertainty, has had a negative effect on manufacturing industries (Fortunato, 2022), which have lost both market share and jobs.

Although the 2010 FIFA World Cup was a boon for the M&E sub-sector, the global financial crisis of 2008/9 hit the local manufacturing sector hard. According to Fortunato (2022), in 2008, local manufacturing experienced "a structural break in its economic trajectory" from which it has yet to recover both in terms of its contribution to GDP and the number of people it employs. He states that manufacturing went from representing 14% of GDP in the first 14 years of democracy (1994–2008) to "making virtually no contribution" in the next decade (2008–2018), in the process accounting for the biggest loss of jobs – or "the largest negative contribution to employment growth" – in the economy.

In 2019, before COVID-19 hit our shores, an estimated 1.8 million people were employed in manufacturing in South Africa; by 2021, the number of people employed in the sector had fallen to 1.4 million. As a result, the sector accounted for less than 12% of employment, down from 17% in 2006, yet it remained the fourth largest sector in terms of its contribution to the country's GDP, at 13%. On the eve of the country's founding democratic elections in 1993, manufacturing was the largest economic sector, accounting for 25% of GDP (Mongalo, 2023).

During the period leading to the 1994 elections, investment, as measured by the GFCF, which had averaged 1.6% in the preceding five years, fell by 11%. This falls way short of the target set in the National Development Plan of 2011, which advocated the creation of close to one million manufacturing jobs over 20 years and set a GFCF target of 30% of the economy by 2030 (Mongalo, 2023).

In the aftermath of the global financial crisis, investment levels declined sharply in the manufacturing sector, with its capital–labour ratio declining. Demand for domestic manufactured goods also plummeted by 18% between 2008 and 2018, equating to 2% lower demand per year, largely because of a fall in both consumption and investment. Concomitantly, there was also a smaller decline in foreign demand for South African manufactured products (Fortunato, 2022).

Although import penetration increased from 1994, following the removal of import tariffs, Fortunato (2022) dismisses the commonly held view that South African manufacturing's woes are wholly attributable to an increased flow of imports into the country. Imports, he states, are merely "a secondary problem". Consequently, he avers, boosting economic growth – rather than import substitution – is the best way of increasing demand for manufactured products.

Before the global financial crisis of 2008, South Africa's longer-term de-industrialisation was consistent with the type of premature de-industrialisation which could be expected in developing countries. However, after 2008, the country's de-industrialisation was "exceptional": "When compared to other developing countries, South Africa had one of the largest falls in both value-added and employment growth rates within manufacturing. During the 10 years following 2008, the fall in manufacturing employment share was the same as during the 30 years preceding 2008," avers Fortunato (2008).

The fall in manufacturing value added and employment growth rates was accompanied by a loss in productive capabilities within the sector. As a result, there was less diversity within the sector. Given that the drop in South African manufactured exports resulting in the decline in foreign demand mentioned above was more acute than the drop in global demand across different manufacturing sub-industries, Fortunato (2008) concludes that South African manufacturing had lost its competitiveness.

In addition to de-industrialisation negatively affecting job creation, South Africa's manufacturing sector also found itself with a lower economic complexity index, which means that over time it lost both the capability to produce more complex products and competitiveness. Not only has South Africa lost the ability to diversity its manufacturing capability, but the diversity of its manufactured exports is now also lower than it was in 1994 (Fortunato, 2022).

Fortunato (2022) argues that instead of Government policy focusing on innovation and technology support to create conditions for companies to diversify into new products and for new manufacturers to enter the market, it has sought to prioritise strengthening industries in the mineral–energy value chain through beneficiation, alleviating job losses in labour-intensive industries, and incentivising local demand through the designation of certain locally manufactured goods. This approach, Fortunato (2022) argues, has focused on the declining number of sub-industries that have historically comprised the manufacturing sector, when future growth and new job creation are likely to be in new industries which align the country's comparative advantages and new opportunities to serve international demand.

The loss of competitiveness and productivity saw South Africa's manufacturing exports losing market share across the board, except in the case of the auto-manufacturing sub-sector. Large export categories such as chemicals and metals lost market share in both the export and domestic markets, and exports in sub-industries like machinery, electronics and textiles declined even as their global markets experienced growth (Fortunato, 2022).

While Bhorat and Rooney (2017) attribute the stagnation of South Africa's manufacturing sector to both the abundance of cheap labour in Asian countries like China, India, Vietnam, and Indonesia and a shortage of skills in the country, Fortunato (2022) contends that local factors are to blame for the country's "exceptional deviation from the premature de-industrialisation curve". The latter is a graph which indicates the pace at which developing countries de-industrialise. He argues strongly that the decline in manufacturing and, generally, South Africa's poor economic performance is caused by the country's lack of reliable electricity supply. According to Fortunato (2022), South Africa's electricity crisis is more directly responsible for the collapse of manufacturing than other factors like "relatively high wage levels" and the Government's localisation policies, such as designation.

The latter is a Government policy which requires state-owned entities and the public sector to purchase locally manufactured goods which the Government considers to be particularly vulnerable to imports. Such interventions often follow intensive lobbying by domestic manufacturers of the product in question.

Fortunato (2022) contends that the electricity crisis has by far the most pronounced impact on the manufacturing sector than any supply-side issues have had. This contention is supported by the findings of the World Bank Enterprise Surveys of 2007 and 2020 to the effect that the percentage of local manufacturing companies which identified electricity as their main production constraint rose from 19% to 62% of the sample, with as many firms using generators to alleviate the challenge as is the case in Nigeria, Liberia and Lebanon.

Compounding the situation is the fact that South African manufacturing's electricity consumption per worker is high relative to its income level. According to Fortunato (2022), the country's energy consumption as a share of GDP is also high compared to that of the rest of the world. In the decade after 2008, electricity became a major constraint for the South African economy in general and for manufacturing in particular, both through substantial price increases and frequent bouts of outages referred to as load-shedding. From 2007 to 2019, average electricity prices more than trebled and load-shedding worsened considerably. Fortunato (2022) concludes that the fact that the number of manufacturing companies owning or sharing generators increased from about 20% in 2007 to above 60% in 2020 means that electricity outages posed a more serious challenge to local manufacturers than anything else, as found by the World Bank Enterprise Surveys.

"This is a particular challenge for manufacturing firms since it is very difficult to maintain high productivity and global competitiveness when powering a significant amount of production through generators. This is especially problematic when fuel prices are high, as is the case in South Africa today", Fortunato (2022) concludes.

Contrary to the argument by some employers' organisations that high labour costs render local manufacturers uncompetitive, Fortunato (2022) argues: "The patterns of manufacturing growth over time and its geographical footprint show relatively high labor costs are not a key constraint for most of South African manufacturing. ... Meanwhile, we find evidence that new tariff protection should be approached with caution based on its misalignment with the cause of the manufacturing decline and clear evidence of negative impacts to downstream industries".

In addition, average wages in the manufacturing sector in South Africa have been found to be higher than those in many other comparable countries, such as Slovakia, and to be closer to those in higher-income countries. The same applies to the wages of managers in the sector, who earn more than their counterparts do in countries like Thailand (Barnes, Black & Techakanont, 2017; Fortunato, 2022). Fortunato (2022) contends that South African manufacturing labour costs "have not increased significantly" at the industry level since 2008 and dismisses the view that they are the driver of local manufacturing's decline or lack of competitiveness. He points out that relatively high labour costs have been "a longer-term feature" of the local economy and that in the past local manufacturing was able to grow "even under this condition".

In the first quarter of 2023, manufacturing in South Africa was a R513-billion industry in terms of gross value added and accounted for 11.2% of GDP, down from 15% of GDP in 1995. It was the fourth largest sector of the economy, though it employed 44 450 fewer people than it did in March 2019, before the onset of the COVID-19 pandemic. However, although production volumes were also 6% less than they were in March 2019, there were indications that the sector was becoming more resilient to the damaging effects of load-shedding (Mhlanga, 2023; Zwane, 2023).

Fortunato (2022) argues that rather than high labour costs being the culprit in the sector's lacklustre performance, it is the Government's designation policy, which is intended to benefit local manufacturers, and the imposition of tariffs on imported goods, which have negative consequences on downstream industries which rely on those products as inputs.

His study found that South Africa's manufacturing is concentrated in the M&E, petrochemical, automotive, and food industries, which collectively accounted for 70% of real output, 74% of fixed capital stock and 65% of exports from the sector in the first 14 years of democracy. The first three of these sub-industries each accounted for about 20% of total manufacturing sales, with the automotive sub-sector representing a smaller share.

## Conclusion

Manufacturing in South Africa owes a lot to the First and the Second World Wars, which saw the then colonial power, Britain, mired in conflict and unable to produce sufficient products for its colonies like South Africa. As a matter of necessity, the South African government of the time set up several industries to encourage and support manufacturing, which was initially based on agriculture and mining. South African manufacturing owes even more to the support that it received from the pre-democracy governments, in the form of various incentives and the imposition of import tariffs.

Although, like other sectors, manufacturing initially benefited handsomely from the democracy dividend, it has subsequently suffered considerably following the democratic Government's decision to remove, without prior warning, the protective import tariffs and the other forms of support which were previously available to the sector. This amounted to throwing an uncompetitive, previously-cloistered sector of the economy to the wolves and expecting it to survive. Over the years, manufacturing has shrunk considerably, both as a source of employment and as a contributor to the country's GDP.

The global financial crisis of 2008/9 made things even worse for South African manufacturing, which has yet to return to pre-2008 levels of capacity utilisation, employment and contribution to the economy. The fact that the mining industry also struggled during the period following the global financial crisis worsened things for the manufacturing sector, especially for its M&E sub-sector, which is a supplier to mining and construction, among other industries.

Although the Government has often touted manufacturing as a potential source of employment for those with low levels of skills, it has not done much to support the sector. Save for the belated imposition of some import tariffs and the designation of some manufactured products for use by the public sector, not much has been done to incentivise manufacturing and to place local manufacturers in a position in which they can be internationally competitive.

It is understandable that South African manufacturing, which has historically been heavily reliant on government support, would clamour for the imposition of more import tariffs and for a more rigorous implementation of designation. Ironically, however, the imposition of import tariffs and localisation through designation are not a panacea for South African manufacturing's problems. Instead of encouraging innovation and improving competitiveness so that local manufacturers will be better placed to take advantage of the opportunities presented by the African Continental Free Trade Area (which seeks to create a massive, barrier-free common market on the continent), the Government has focused on interventions which have the effect of making local manufacturers more sheltered and less competitive, both domestically and internationally.

# Chapter 2

# Understanding the Metals and Engineering Sub-Sector

## Introduction

The M&E sub-sector is a very important part of the broader manufacturing sector. Other sub-industries constituting the manufacturing sector are petroleum and chemical products, rubber and plastic products, basic iron and steel, non-ferrous metal products, metals products and machinery, electrical machinery, motor vehicles and accessories and other transport equipment. Over the years, the M&E sub-sector has constituted a large part of the manufacturing sector, and there have been times when it was the largest industry within manufacturing.

This being the case, the trajectory of the sub-sector's development in the country has broadly mirrored that of the manufacturing sector, as was explained in the previous chapter. In fact, Tafadzwa Chibanguza, Chief Operations Officer of the Steel and Engineering Industries Federation of Southern Africa (SEIFSA), describes the M&E sub-sector as a coincident economic indicator, meaning that it is so important that its performance tends to mirror that of the economy in general (Chibanguza, 2023).

The M&E sub-sector itself is quite broad and very diverse. Its constituents range from primary steel producers through to fabricated metal producers. Its components are listed below, together with their respective weightings within the sub-sector for 2018–19, as calculated by Statistics South Africa (Ade, 2019):

- Rubber products (4.62%)
- Plastic products (9.86%)
- Basic iron and steel products (11.86%)
- Basic non-ferrous metals products (9.41%)

- Structural metal products (6.41%)
- Other fabricated metal products (13.31%)
- General-purpose machinery (8.66%)
- Special-purpose machinery (12.01%)
- Household appliances (2.79%)
- Electrical machinery and apparatus (5.69%)
- Bodies for motor vehicles, trailers and semi-trailers (1.59%)
- Motor-vehicle parts and accessories (9.62%)
- Other transport equipment (4.17%).

Obviously, the picture changes from time to time. At the beginning of 2023, when the sub-sector represented 26.15% of manufacturing, the composition of its constituents was in Table 2.1 (Chibanguza, 2023).

**Table 2.1**: Sub-industries within the M&E sub-sector. Source: Chibanguza (2023): "SEIFSA State of the Metals and Engineering Sector 2023"

| Sub-Sector | % of Manfacturing | M&E Weights (% |
|---|---|---|
| Plastic products | 2,29% | 8,8% |
| Basic iron and steel products | 2,82% | 10,8% |
| Non-ferrous metal products | 3,26% | 12,5% |
| Structural metal products | 1,98% | 7,6% |
| Other fabricated metal products | 3,35% | 12,8% |
| General purpose machinery | 3,46% | 13,2% |
| Special purpose machinery | 3,87% | 14,8% |
| Household Appliances | 0,73% | 2,8% |
| Electrical machinery and apparatus | 2,31% | 8,8% |
| Bodies for motor vehicles, trailers and semi-trailers | 0,71% | 2,7% |
| Other transport equipment | 1,37% | 5,2% |
| Total M & E Sector | 26,15% | 100,0% |

## The Rise of the M&E Sub-Sector in South Africa

Like the rest of the manufacturing sector, the M&E sub-sector owes much gratitude to General Jan Smuts and Dr Hendrik van der Bijl, who were responsible for the establishment of Eskom and Iscor to provide cheap electricity and steel to local manufacturers (Jacobs, 1948).

Long before that, the discovery of gold and the expansion of gold mines in South Africa in the 1880s had spurred demand for iron and steel in the country. To meet that demand, iron merchants imported steel from Europe. The second steel plant to be established in South Africa was the short-lived Transvaal Government Iron Concession, which was built by the government of the Transvaal Republic to meet demand for steel that would be needed to build a railway line between Lourenço Marques (now Maputo) and the Transvaal. Upon completion of that project, the company was liquidated because investors' attention had turned to the more lucrative gold mines (Dondofema, Matope & Akdogan, 2017:2).

Before the Transvaal Government Iron Concession, the South African Coal and Iron Mining Company had been formed in 1882 in the Dundee region of Natal (now KwaZulu-Natal). It, too, did not last long. Next to be formed, in 1912, was the Union Steel Corporation (USCO) of South Africa Limited, whose focus was the treatment of railway scrap metal. South African Railways was required by law to buy its iron and steel from USCO (Dondofema *et al.*, 2017:2).

As was the case with the broader manufacturing sector, M&E industries also benefitted handsomely from the First and Second World Wars. When the First World War began in 1914, all steel imports were restricted, much to the benefit of local independent steel companies like Dunswart Iron and Steel Works, which had begun its life in Benoni as Cartwright and Eaton Machine Merchants and, in 1911, was renamed Union Iron and Steel Works. Among the immediate beneficiaries of the imposition of steel import tariffs was the Witwatersrand Cooperative Smelting Works, which was established by the Robinson Gold Mine (Dondofema *et al.*, 2017:4).

A few other relatively small steel companies, which did not last beyond the apartheid era, were subsequently formed. They had progressively higher levels of production capacity as a result of the installation of modern blast furnaces. In 1924, a big player which was to outlast the apartheid era, Scaw Metals, was formed in Eloff Street Extension in Johannesburg to cater for the railway market. It subsequently acquired a 49-hectare site in Germiston, to which it relocated, and, by 1956, it was the main supplier of cast steel bogeys (Dondofema *et al.*, 2017:5).

Another significant player was Pretoria Iron Mine, which was formed by a Mr Delfos, who built an experimental blast furnace in 1917. Although Pretoria Iron Mine ceased production towards the end of 1919, it continued with exploration in the iron and steel market, and subsequently managed to supply up to 50% of South African Railways's projected iron and steel requirements. To ensure that it could meet this requirement, Mr Delfos entered into a partnership with a number of organisations, among them Anglo American and the National Industrial Corporation (which was later to become the Industrial Development Corporation), to form the South African Iron and Steel Corporation. However, this immediately led to a complaint by USCO, which had the right to treat railways scrap metal (Dondofema *et al.*, 2017:5). A government committee appointed to investigate USCO's complaint concluded that the agreement by the various stakeholders to form the South African Iron and Steel Corporation was legally invalid.

South Africa's behemoth steel producer, Iscor, was formed in June 1928, following the promulgation of the Iron and Steel Industry Act in the same year. It was initially meant to be a partnership between the government and the private sector, with the latter meant to contribute three million British pounds towards its 3.5 million pounds capital structure. However, when the private sector did not come to the party, the government acquired the entity's B shares and became Iscor's primary owner (Dondofema *et al.*, 2017).

South Africa's declaration of war on Nazi Germany in 1939, which saw the country enter the Second World War, led to higher demand for locally produced iron and steel. That created opportunities for more local players to enter the industry, with others situating themselves in the value chain as importers (SEIFSA, n.d.).

Until then, South Africa's M&E sub-sector was focused on repair and maintenance work. However, the country's entry into the war led to "massive expansion and technological development" in the sub-sector. As happened to the broader manufacturing sector during the First and Second World Wars, the local economy found it difficult to source M&E production equipment and finished products from abroad. This saw the fledgling local M&E sub-sector rising to the challenge "with remarkable ingenuity and resourcefulness" to meet the growing need. Within a space of a few months, South African engineering workshops were converted into "efficient munitions factories" and started producing a growing range of weaponry and spares comparable in quality with weaponry produced in the UK or the USA (SEIFSA, n.d.).

Consequently, the Smuts government sourced most of its war-time requirements from South African operations. According to SEIFSA (n.d.), during the war years (from 1940 to 1944), South African M&E companies manufactured and delivered:

- More than 57 000 armoured fighting vehicles and 35 000 military motor vehicles;
- Over 70 000 tonnes of bombs, 50 000 tonnes of shells and shell casings, and 28 000 tonnes of mortar bombs, grenades and landmines;
- Nearly 20 000 tonnes of small arms and ammunition; and
- Over 1.5 million helmets.

Other players subsequently entered the industry, such as African Metals Corporation Limited in Newcastle and the Minerals Engineering Company of South Africa in Witbank, owned by the Rockefeller family of the USA. In later years, Anglo American purchased two-thirds of the Minerals Engineering Company and renamed it the Transvaal Vanadium Company Limited. In 1960, the Highveld Steel Development Company was established (Dondofema *et al.*, 2017).

Among other worthy developments in the sub-sector were that

- Cape Gate (Pty) Ltd was established in 1962;
- Cape Town Iron and Steel Works (Pty) Ltd was established in 1965, but was taken over by Iscor eight years later; and
- In 2005, Unica Iron and Steel (Pty) Ltd began production.

Says SEIFSA (n.d.) in its articulation of the sub-sector's history:

> Out of these wartime facilities came aircraft hangars, bridges, travelling cranes, barges, rock drills, pumps and valves, electric motors, transformers, field hospital equipment, ships' stores for the British Admiralty, thousands of kilometres of telephones and telegraph wire for the signal services ... the list is almost endless.

> The Western Desert pipeline supplying water for allied troops in North Africa was built largely from South African metals.

Clearly, then, the Second World War was very good for the M&E sub-sector in South Africa, just as it had been for the broader manufacturing sector in general. However, the end of the war did not presage the beginning of a slowdown in the M&E sub-sector. Instead, the end of the Second World War led to "an explosive increase" in these industries. More players entered in the second half of the 1940s, leading to the number of companies in it growing "by many hundreds" and the number of people employed in the sub-sector rising from 70 000 to 110 000 (SEIFSA, n.d.).

Inevitably, this flurry of activity led to growth in the sub-sector's manufacturing complexity index. For the very first time, domestic appliances and radios were manufactured in the country, and a company called Brown Boveri Technologies started producing electrical power and control equipment. Considerable investments were also made in the production of copper and copper alloy and semi-fabricated aluminium products. This led to the government introducing the South African Bureau of Standards to monitor the quality of production standards and performance specifications for local manufacturers (SEIFSA, n.d.).

Although many economists had forecast a global recession following the end of the Second World War, as had occurred after the First World War, the South African economy exhibited great resilience. Thanks to a myriad of factors, such as international collaboration, monetary controls, lease lending and financial aid to developing countries, South Africa experienced "a fairly smooth transition from war to a peace-time economy" in the

1950s and 1960s. As it continued to recover, post-war South Africa – with growing purchasing power and a need for capital equipment and plants, among other things – continued to import whatever it did not manufacture locally. As a negative balance of payments ensued, the government introduce "drastic exchange control measures" (SEIFSA, n.d.).

However, those measures did not remain drastic for all time. When the country started earning substantial foreign exchange from gold mining and uranium exports from new gold fields in the Orange Free State, the foreign exchange deficit moderated and the exchange controls were eased. The local M&E sub-sector continued to grow and to show even higher levels of sophistication. As SEIFSA President RM Russel noted in his address to the organisation's annual general meeting in 1950, South African M&E companies had expanded their production capacity, some had concluded partnerships with foreign counterparts and new players had entered the sub-sector "to produce a surprisingly wide variety of articles" which until then had been imported (SEIFSA, n.d.).

As the local M&E sub-sector became more internationally competitive, some of its players started exporting to foreign markets. The advent of the Korean War in the 1950s led to a shortage of many commodities on the global markets, and this led to a rise in the prices of those commodities. Not only did the local M&E sub-sector survive the Korean War, but it grew even further and saw more new companies entering its sub-industries. Consequently, the number of people employed in the sector grew to more than 140 000 and its output soared (SEIFSA, n.d.).

The government's imposition of import restrictions in 1949–50 led to the opening of the second Iscor steel works, at Vanderbijlpark, in 1952, and by the end of that decade Iscor's production output had risen to two million tonnes. Diversification by some mining companies, which invested in M&E entities, saw an increase in the number of licensing and trade know-how agreements concluded with international counterparts (SEIFSA, n.d.).

South Africa, which had become a whites-only republic in 1948 and more viciously oppressed black people (Africans, coloureds, and Indians), was ill prepared to deal with the various controversies in which it found itself mired in the 1960s as the country because more politically fragile and the government was internationally ostracised. While some white South Africans emigrated during that decade and property prices plummeted, many others and the government doubled down and hoped to survive international isolation. To make a success of "Fortress South Africa", the apartheid government became more generous in its support of local manufacturers, including those in the strategic M&E sub-sector of the economy. The aggressive import substitution efforts paid off handsomely as the economy boomed and the country's GDP "rocketed from R5 billion in 1960 to R13 billion in 1970" (SEIFSA, n.d.).

SEIFSA (n.d.) argues in its history that the long-term import substitution strategy embarked upon by the government at the time and the infrastructure development which took place "still benefit South Africa today". As an example, the organisation cites the "booming automotive vehicle and components manufacturing industry", which was conceived in 1960 when the then Minister of Economic Affairs enjoined the Board of Trade and Industries to consider ways of boosting local content production.

"The 'Build the Economy' policy brought unprecedented spin-offs for the metal and engineering industry, which forged ahead at an average 12% during the mid-Sixties. The workforce burgeoned to 265 000," observed SEIFSA (n.d.) triumphantly.

So advanced was the sub-sector that, by the end of 1966, it employed 28 000 artisans and 6 000 apprentices (SEIFSA, n.d.).

In response to punitive economic sanctions in the 1980s, the apartheid government invested in ambitious infrastructure and import-replacement projects, offered local manufacturers generous export incentives, and imposed even harsher tariffs on imports. The M&E sub-sector was among the biggest beneficiaries of those efforts and accounted for up to a third of South Africa's manufacturing, valued as it was at more than R10 billion (SEIFSA, n.d.).

However, the severe economic recession of 1981/82 which began in the USA was also felt in South Africa. By 1983, the country's M&E sub-sector was experiencing an economic depression that was the worst since the Great Depression half a century earlier (SEIFSA, n.d.).

Iscor was privatised in 1989, half a decade prior to the attainment of South Africa's democracy, and listed on the Johannesburg Stock Exchange. Two years later, it took control of USCO and renamed the facility Iscor Vereeniging Works. In the democratic era, Iscor's ownership changed first to Kumba Resources Limited, and later LNM Holdings owned 51% of it. Following the merger of Mittal Steel with Arcelor in June 2006, Iscor was renamed ArcelorMittal South Africa (Dondofema *et al.*, 2017).

As the punitive economic sanctions imposed on apartheid South Africa in the 1980s began to bite, the economy took a massive beating as some foreign companies disinvested. Consequently, investment in the local economy took a dive from an average of 30% of the GDP in the early 1980s to a paltry 16% by the early 2000s. In its diagnostic analysis, South Africa's National Development Plan 2030 was to conclude that the country had lost out on "a generation of capital investment in roads, rail, ports, electricity, water, sanitation, public transport and housing" (Ryan, 2022).

The overnight removal of the high import tariffs after 1994, which had protected the sub-sector until the dawn of democracy, left the local M&E sub-sector exposed to international competition at a time when it was not ready for it. Lower commodity prices from 2006 onwards made the situation even worse for the South African iron and steel industry. Consequently, local steel production fell from a high of 9.7 million metric tonnes in 2006 to 6.55 million metric tonnes in 2014 (Dondofema *et al.* 2017:8). Imports of iron and steel more than doubled in four years: in 2006, South Africa imported 2.8 million metric tonnes of iron and steel, but by 2015 the country was importing as much as 5.98 million metric tonnes of iron and steel, marking a 113% increase in imports. This had a devastating impact on the local iron and steel industry: Cape Town Iron and Steel Works stopped production in 2009; ArcelorMittal South Africa's Vanderbijlpark mini-mill plant closed in 2012; the company's Vereeniging mini-mill plant closed three years later; and EVRAZ-Highveld Steel Vanadium Corporation shut down in 2016 (Dondofema *et al.*, 2017:8−9).

## Global Developments in the M&E Sub-Sector

The end of the Second World War marked the beginning of a great period for the global M&E sub-sector. The major infrastructure developments which took place in Europe and elsewhere after the Second World War presented a boon to the M&E sub-sector, which was pivotal in most forms of development. In particular, the M&E sub-sector in the USA, the country which had emerged victorious from that war, enjoyed such technological superiority that America became the world's unrivalled steel producer. According to Kenward (1987:30), by 1955 the world sourced 40% of its steel needs from the USA and exported to that country a mere 1% of its domestic steel needs. So large was the sub-sector in the USA at the time that it employed up to 1% of the working civilian population of the country (Kenward, 1987:30). The main reason for this state of affairs was that Germany's and Japan's steel-making capacity was destroyed during World War II (Coffin, 2003:2).

In northwest Louisiana, almost 30% of the local workforce was employed in steel mills and the total value of the steel output of the USA in 1969 was $70.9 billion, at a time when the country's GDP was $3 492 billion (Coffin, 2003:1).

By 1953, the M&E sub-sector in the USA employed 650 000 people, the highest it achieved, though that number fell to 512 000 by 1974, 399 000 by 1980, and 236 000 by 1994. US steel production peaked at 111.1 million tonnes in 1973 and declined to 97.9 million tonnes in 1978 and 70 million tonnes by 1984. The next peak in steel production occurred in 2000 when 100 million tonnes were produced, before it dropped to 86 million tonnes in 2014. Steel imports into the US rose from 146 000 tonnes in 1946 to 24 million tonnes in 1978 and, eventually, to 34.5 million tonnes in 2017 (Encyclopedia, Science News and Research Reviews, n.d.).

From 1950 to 1970, steel production across the world accelerated at a phenomenal pace, though this occurred at a much slower pace in the USA. For instance, while it grew by an estimated 16% per annum in Japan during those two decades, 10% per annum in developing countries, and 5.5% per annum in the European Community (now the European Union), steel production grew by a more sedentary 1.5% per annum in the USA. Concerned about the amount of steel imported into the country, in 1968 the US government concluded three-year, voluntary restraint agreements with the Federal Republic of Germany and Japan; these agreements were subsequently extended in a modified form for another three years (Kenward, 1987:30).

By 1972, the M&E sub-sector in the USA had lost so much competitive advantage that its unit operating costs were about 40% higher than those of its primary competitors in Japan, where labour-productivity levels were higher and the cost of iron lower. Wages in this sub-sector in the USA were almost 50% higher than the average for manufacturing overall in the country at the time (Kenward, 1987:32). Tarr (1988:183) adds that between the 1960s and the early 1970s, steelworkers around the world generally earned about 50% more than their counterparts in other manufacturing jobs, and that, by 1982, US steelworkers earned as much as "93 percent more than the US average".

However, for the sake of context, it is important to keep in mind that between 1948 and 1973, the American economy was thriving and the country's gross national product averaged 3.7%, that unemployment was low, and that, at its highest, the inflation rate was around 5%. The world's manufacturer at the time, the USA was a net exporter and enjoyed a trade surplus of $157 billion (Whitney, 1987).

All of that changed during the early 1970s when a confluence of factors adversely affected the US economy. Following President Jimmy Carter's October 1973 request that the US Congress set aside $2.2 billion aid to supply arms to Israel for the Fourth Arab–Israeli War of that year, Arab members of the Organization of Petroleum Exporting Countries placed an oil embargo on the US and other Israeli-supporting countries like Netherlands, South

Africa and Portugal. The resulting oil shortage led to petrol prices at filling stations soaring in those countries. In the US, the oil price quadrupled from $2.90 per barrel to $11.65 by January 1973 (US Department of State, n.d.; Corbertt, n.d.).

Consequently, the prime rate in the US rose to 20% in 1981, before finally falling to 8% in 1986, and GDP growth averaged 2.3% between 1973 and 1985. In the latter year, the US's net export deficit was $79 billion, which was equal to half the combined trade surpluses the country had achieved between 1948 and 1973 (Whitney, 1987). From 1979 to 1982, more than 150 000 people who had been employed in the steel industry in the USA lost their jobs; the number for the decade 1976–1986 was double that as several steel mills shut down (Rowe, 2016).

A global steel crisis occurred in 1973–1975, which saw steel prices plummeting because of oversupply of the commodity. Consequently, many steel mills went out of business in North America, the UK, West Germany, and Sweden (Encyclopedia, Science News and Research Reviews, n.d.; Coffin, 2003:3).

In the UK, employment levels in the steel industry fell from 197 000 in 1974 to 179 000 in 1977, 112 000 in 1980, and fewer than 62 000 in 1984. In the then nine-member European Community, steel production trebled between 1950 and 1970, with the community remaining a net steel exporter until the 1980s. The number of employees in the industry in the European Community fell from 795 000 in 1972 to 722 000 in 1977, 598 000 in 1980 and, eventually, to 440 000 in 1984 (Encyclopedia, Science News and Research Reviews, n.d.).

A second – and more severe – recession in the USA in 1981–82 adversely affected that country's economy. The unemployment rate rose to 7.5%. US manufacturing had reached its all-time peak in employment in June 1979; a year later manufacturing companies had to let go of 1.4 million employees. Altogether, 10% of jobs (2.3 million) in manufacturing were lost during this period (Plunkert, 1990:9; Sablik, n.d.). By the late 1980s, the US was importing up to a quarter of its steel needs (Kenward, 1987:33).

Among the countries which benefitted from the reduction in steel production in the USA was China, which in 2011 emerged as the largest steel producer in the world, accounting for 45% of international production; by 2020 it was producing more than half of the world's steel. On its own, China produced more steel than the next four largest steel-producing countries (MacKay, 2023). By far the largest steel producer in the world, China is also the largest consumer of the material (Seth, 2022).

## South Africa's M&E Sub-Sector in the 21st Century

The M&E sub-sector of manufacturing is a strategically important component of the South African economy. A country requires a competitive M&E sub-sector, or relatively good access to its' products, in order to achieve a developed infrastructure.

As indicated earlier in this chapter, over the years, the performance of South Africa's M&E sub-sector has mirrored that of the economy. Its performance has been cyclical, thus making it "a coincident economic indicator" (Ade & Kruger, 2019; Chibanguza, 2023). The country's economy averaged a growth rate of about two percentage points per year between the year 2000 and 2017, and the M&E sub-sector expanded by an average of 16% during that period. The economy had its best performance between 2003 and 2006 when it grew by an average of 4.6% per annum, though it subsequently averaged 2.7% between 2007 and 2009, 2.8% between 2010 and 2013 and 1% between 2014 and 2017. During the same periods, the M&E sub-sector shrunk by an average of 4%, rebounded to 2.1%, and then registered 1.3% growth in 2014–17 (Ade & Kruger, 2019:54).

During the same broad period (2000 to 2016), the M&E sub-sector exported a higher percentage of its products, especially machinery and equipment, while there was volatility in the amount of imports into the associated sub-industries. In 2019, 52% of the sub-sector's products were exported. Between 2001 and 2008, most of the sub-sector's exports headed to Europe, but that changed briefly the following year when Asia became the primary export destination. However, from 2010 onwards, Africa emerged as the sub-sector's largest export market, with 37.2% of its products headed for other countries on the continent. Asia became the second-largest export market (26.6%), followed by

Europe in third place (22%), the Americas in fourth place (12.3%) and Oceania in last place (1.6%), with 0.3% of exports unaccounted for (Ade & Kruger, 2019:97). In 2018, South Africa's M&E sub-sector had its best trade balance (R80 billion) with Africa.

The local market accounts for 48% of the domestic M&E sub-sector's products. Its local market includes the mining (9.7%), automotive (12.7%), construction (14.6%) and petrochemical industries, and some of its products are consumed within itself. Its products are mostly intermediate goods which are used as inputs in subsequent production processes. The M&E sub-sector itself primarily procures its inputs from the mining (41.7%), petrochemical (4.1%) and agriculture, forestry, and fisheries industries. In addition to the backward and forward linkages with other industries, the M&E sub-industries also have a fair amount of lateral linkages (Ade & Kruger, 2019:57−8).

The 2008/9 global financial crisis (GFC) marked a major turning point in the fortunes of the manufacturing sector in South Africa. Before 2008, the M&E sub-sector was responsible for the largest growth in employment in the economy; following the crisis, it was revealed to have been the largest contributor to job losses. Up to 45% of the total number of jobs lost in the manufacturing sector since 2008 occurred in the M&E sub-sector (Bhorat & Rooney, 2017; Fortunato, 2022). Although jobs were lost across almost all M&E sub-industries, the worst affected were the labour-intensive ones such as fabricated metal products, structural metal products, and special-purpose machinery. Subsequently, profit margins in the sub-sector came under immense pressure (Fortunato, 2022).

Chibanguza (2023) characterised the M&E sub-sector as having been "locked in a deep, multi-year structural recession" since the GFC of 2008. For instance, between 2008 and 2015, 34 500 jobs were lost in fabricated metal products, at an annual rate of 6%, and 21 000 jobs were lost in structural metals, at an annual rate of 4% (Fortunato, 2022). The sub-sector recorded its highest production figure of 120 index points in 2002. Relative to pre-crisis levels, production declined by 15% between 2007 and 2009, by 16% between 2010 and 2013, and by 18% between 2013 and 2016 (Ade & Kruger, 2019:51). By 2016, the sub-sector was in a recession, thanks to a cumulative revised contraction of 7.3% between 2013 and 2016 because of a decline in investment over a 23-year period (Ade & Kruger 2019:49).

In response to calls from industry players (such as NUMSA, ArcelorMittal, SEIFSA and others), in 2015 the Government imposed economic tariffs to protect the country's primary steel producers from dumping. ArcelorMittal South Africa, which had not recorded a profit in five years (and whose share price had declined by 60% in a year), offered to conclude a black economic empowerment (BEE) deal if the Government responded positively to its request for the imposition of steel-import tariffs. The general rate of customs duty on primary steel products was raised to 10%, and a safeguard duty on hot-rolled coiled and plate products was imposed (SEIFSA, n.d.)

These tariffs proved to be very unpopular among a number of players, especially metal fabricators, who relied on the primary steel producers. As Fortunato (2022:36) subsequently pointed out seven years later, it is easy for high tariffs which are intended to protect one M&E sub-industry to "create major challenges for downstream industries that pay the cost of those tariffs." This policy response, Fortunato (2022) argues, did not target the causes of the collapse among primary steel producers; hence, it may have been more deleterious to the M&E sub-sector. Indeed, Ade and Kruger (2019:74) also cautioned that the imposition of import tariffs could only serve as a temporary solution. Instead, they added, it would be vital for the sub-sector to be internationally competitive both at home and abroad in order for its sub-industries to emerge as "preferred suppliers domestically and in international markets".

The year 2017 proved to be slightly better for the sub-sector, which registered an improved growth rate of 2.7% year on year (Ade & Kruger, 2019:50). During that year, the M&E sub-sector recorded a cumulative trade deficit of R121.5 billion: it exported goods valued at R237.2 billion and imported goods worth R358.7 billion. A year earlier, the sub-sector's trade deficit had improved by 15%. The biggest imports were machinery equipment (66.1%)

and base metals (15%), with the remaining sub-sectors being plastic and rubber, and vehicles and transport, representing 12.3% and 6.4%, respectively. The imports into the sub-sector originated from Asia (51.9%), Europe (33.4%), and the Americas (10.9%) (Ade & Kruger, 2019:101). Ade and Kruger (2019:81) also made the point that, over a 14-year period (between 2005 and 2019), labour productivity in the M&E sub-sector had been disappointing. While labour productivity had improved by 17%, the unit cost of labour had increased by 54% during the same period.

"The situation is not sustainable," Ade and Kruger (2019:81) cautioned. "The cost of each successive wage increase is not only measured in terms of the percentage increase in the wage, but it is compounded by stagnant labour productivity."

In the first quarter of 2017, the M&E sub-sector represented 29.17% of the manufacturing sector and contributed 3.5% to the country's GDP. Five years later, it represented an even smaller percentage of manufacturing, namely 26.15%, and contributed 2.5% towards South Africa's GDP (Gumede, 2023). In the same year (2022), real gross domestic fixed investment in the M&E sub-sector contracted by -9.2%, and a further contraction of -1.5%, in a best-case scenario, was expected in 2023, with the worst-case scenario expected to be a contraction of -2.2%. This situation was attributable to South Africa's then worsening electricity supply outages (Chibanguza, 2023).

Table 2.2 presents a view of the M&E sub-sector at a glance.

**Table 2.2:** The M&E sub-sector at a glance. Source: "SEIFSA Annual Review" (2023)

| METALS AND ENGINEERING SECTOR ECONOMIC DASHBOARD | | | |
|---|---|---|---|
| Economic Variable | 2020 | 2021 | 2022 |
| M&E Production (% growth/contraction) | 12.3% | 27.2% | 1.8% |
| M&E GDP (Rand billion) | 118.7 | 129.6 | 131.1 |
| M&E GDP (% growth/contraction) (2015 Prices) | -23.9% | -0.9% | 6.8% |
| Manufacturing Sector Share of GDP (%) | 15.0% | 15.6% | 14.8% |
| M&E Share of Manufacturing (%) | 17.0% | 17.1% | 17% |
| M&E Share of GDP (%) | 2.5% | 2.6% | 2.5% |
| M&E Capacity Utilisation (%) | 66.6% | 75.3% | 75.6% |
| M&E Sector Input Cost Inflation (%) | 12.4% | 9.8% | 13.1% |
| M&E Employment (number) | 371 956 | 371 396 | 374 496 |
| M&E Employment (% growth/contraction) | -5.1% | -0.2% | 0.8% |
| Gross Earnings (Rand billion) | 96.8 | 105.5 | 107.7 |
| M&E Total Sales (Rand billion) | 638.5 | 808.4 | 814.2 |
| M&E Export Sales (Rand billion) | 256.1 | 323.5 | 342.9 |
| Export sales % of total sales | 40.1% | 40.0% | 37.9% |
| M&E Imports (Rand billion) | 347.4 | 418.9 | 471.1 |
| M&E Trade Balance (Rand billion) | -91.3 | -87.1 | -128.2 |

## Conclusion

Like the broader manufacturing sector, which benefitted immensely from the paucity of imports as a result of the First and Second World Wars, South Africa's M&E sub-sector thrived because of generous government support over the years. It was also a generous beneficiary of the apartheid government's commitment to an armament industry as it sought to be self-sustainable in the face of economic sanctions by many in the international community.

Ironically, like the broader manufacturing sector, the previously cloistered M&E sub-sector found itself prematurely – and harshly – exposed to global competition following the dawn of democracy in South Africa in 1994. In the Uruguay round of multilateral negotiations which culminated in the General Agreement on Tariffs and Trade in 1994, which preceded the formation of the World Trade Organization a year later, a democratic South Africa introduced significant changes to its trade policy (International Trade Administration Commission of South Africa, n.d.): import tariffs were reduced from an average of more than 20% in 1994 to an average of 7.1% in 2020 (International Trade Administration of South Africa, 2023).

That sudden removal of tariffs that had long protected South African manufacturing had a major impact on the country's manufacturing sector, including its M&E sub-sector. Among the worst affected was the clothing, textile and leather sub-industry, but all of manufacturing was adversely affected to varying degrees following the dawn of democracy. This is clearly evident in the manufacturing sector's contribution to the GDP before and after democracy: in 1992 and 1993, respectively, the manufacturing sector contributed 22.31% and 21.39%, but by 2022, this had dropped to almost half (12.04%) (Macrotrends, n.d.).

# Chapter 3

# The History of the Steel and Engineering Industries Federation of Southern Africa

## Introduction

The Steel and Engineering Industries Federation of Southern Africa is one of the oldest employers' organisations in South Africa. Its origin dates back to the beginning of the Second World War, a period that was a boon to South African manufacturing. As Europe and the USA focused their manufacturing efforts on the war, developing countries like South Africa struggled to obtain their usual imports from the West. This had the effect of encouraging local entrepreneurs, who enjoyed great support from the government of the time, to engage in local manufacturing as a substitute for scarce imports.

Given the absence of an independently researched history of the organisation, much of the content of this chapter relies on SEIFSA's own curation of its history and, for the period between 2013 and 2021, on notes and other documents that I made and kept. Unless otherwise stated, the source of the information which follows is SEIFSA, through its archives.

## The Early Years as the South African Federation of Engineering and Metallurgical Associations

In September 1939, South Africa entered the Second World War (WW II) by declaring war on Adolf Hitler's Germany. With a higher domestic demand on the country's then-fledgling M&E sub-sector, which was a vital part of the war effort, it became imperative for employers in the industry to be better organised so that they could speak with one voice in their dealings with both the government and labour. At the time, the former had appointed a Director-General of War Supplies and a Controller of Industrial Supplies, both of whom would liaise closely with the M&E sub-sector. There were also growing concerns among the five existing trade unions, which represented white workers, about the long hours their members worked at a time when there was a wage freeze in the industry.

Until then, employers in the M&E sub-sector, whose work and products were "crucial to the war effort", were organised regionally, hence they could not speak with one voice at the national level. To address this challenge, they held consultative conferences in Bloemfontein and Cape Town in 1941 and 1942, respectively, and finally resolved to form "a national umbrella organisation", called the South African Federation of Engineering and Metallurgical Associations (SAFEMA). The organisation officially began its life in 1943, and later changed its name to the Steel and Engineering Industries Federation of South Africa (SEIFSA).

HC Gearing, from the Cape Engineers and Founders Association (CEFA), which had been established twenty-three years earlier, was elected the founding President of the Federation and former journalist Fred Williams was appointed as the organisation's first Executive Director. To ensure that SEIFSA truly represented all the regions' interests, a SEIFSA Council was formed, which was made up of three delegates and three alternative delegates from the regions, namely the Cape, the Transvaal, Natal, the Natal Midlands, and the Eastern Cape Border areas. At the time, SEIFSA represented a total of 440 employers.

While WW II raged on, South Africa's M&E sub-sector, which had largely focused on repair and maintenance work before then, experienced "vigorous growth". Thanks to the demands resulting from the war, the industry experienced "massive expansion and technological development". Consequently, several new sub-industry-specific employers' associations were formed. These associations took membership of the new federation and subsequently constituted the majority within it. This led to a revision of SEIFSA's original constitution as the new sub-industry-specific employers' associations took their place on the SEIFSA Council, alongside the founding regional founders. The SEIFSA Council had the authority to co-opt, as members, "prominent industrialists whose knowledge and experience would be of value to the federation".

As was the case with the manufacturing sector in general, the WW II was good for South Africa's M&E sub-sector. SEIFSA was to remark gushingly in its curated history:

> Faced with a cut-off in the overseas supply of production equipment and finished articles, [the] South African industry rose to the challenge with remarkable ingenuity and resourcefulness. Within a few months, engineering workshops were converted into efficient munitions factories and began turning out an ever-increasing range of wartime equipment and spares comparable in quality with munitions produced in Britain and the United States.

Among the items produced by the aforementioned wartime facilities, as given by SEIFSA, were aircraft hangars, bridges, travelling cranes, barges, rock drills, pumps and valves, electric motors, transformers, field-hospital equipment, ships' stores, as well as thousands of kilometres of telephone and telegraph wire for signal services.

A year after SEIFSA's formation, in 1944, a national collective bargaining council for the M&E sub-sector, the National Industrial Council for the Iron, Steel, Engineering and Metallurgical Industries, was formed. It concluded the first national wage agreement in the same year.

The M&E sub-sector continued to thrive long after the end of the Second World War as South Africa's economy made a smooth transition from war to peace. The sub-sector developed an impressive level of manufacturing complexity as it transitioned to manufacturing products which had previously been imported. In the 1950s, relations between employers, as represented by SEIFSA, and the labour unions were "amicable and constructive", in SEIFSA's words, thanks to the existence of the National Industrial Council. One of the products of the "amicable and constructive" relations between employers and labour was the formation, in the mid-1950s, of the industry's Sick Pay Fund and the Group Life and Provident Fund for skilled workers. At that time, the South African government had tough legislative restrictions in place for black workers, and SEIFSA's work did not involve engaging with black workers.

In 1953, the National Party government passed the Black Labour Relations Regulation Act, which excluded black people from participation in any form of collective bargaining. Instead, black workers' interests were to be dealt with through a white-controlled central board. In addition to providing for centralised collective bargaining between employers and unions representing white workers, the Industrial Conciliation Act also entrenched discrimination in the workplace. Expressly, that legislation made it illegal for black workers to form a trade union of their own and forbade the formation of racially mixed trade unions.

These two pieces of legislation were buttressed by the Group Areas Act of 1950 and the Influx Control Act of 1952, which, respectively, confined black people to homelands and severely restricted their movement in the country's towns and cities. Before the end of that decade, the government amended the National Conciliation Act to introduce job reservation to protect white workers from competition from black people.

In the 1960s, as some countries began to isolate apartheid South Africa, local businesses became resilient and innovative to survive. They also benefitted from the government's efforts to shelter them from international competition and to promote local manufacturing. As the M&E sub-industry grew, so, too, did SEIFSA, which by 1966 had 41 independent employer associations as members, collectively representing more than 1 600 companies. The growth in the number of member associations, which had different requirements, made it imperative for SEIFSA to set up different divisions specialising in administration, collective bargaining and related labour matters, economics, as well as skills development and training.

In 1965, SEIFSA and its labour partners launched the Metal Industries Medical Aid Fund for skilled employees and, in 1966, the Metal Industries Group Pension Fund as a non-contributory scheme for unskilled employees.

SEIFSA contends that, as the government erected more barriers for black workers in the 1970s, "year after year we [SEIFSA] campaigned for changes to this legislation". As an example, the federation mentions its concerns about the Physical Planning and Utilisation of Resources Act, which limited the number of black workers in specified areas without the permission of the Minister of Planning. It states in its narration of its history that it "called for the Act to be applied in a more flexible manner as it was placing artificial restraints on the mobility and recruitment of black labour".

As its members struggled to fill vacancies because of skills shortages, SEIFSA was involved in the recruitment and screening of white immigrants seeking to be employed in the M&E sub-sector. Since the number of white immigrants attracted into the sub-sector was not enough for its needs, in 1972 SEIFSA reached an agreement with its counterpart trade unions to allow black workers to occupy some jobs previously reserved for whites only. This, says the federation, was "to become an evolving trend". In 1976, that dispensation was extended to include positions just below the level of artisans.

In 1977, the government set up the Wiehahn Commission, the purpose of which was to advise on ways to stabilise labour relations and to facilitate economic growth. Then SEIFSA Executive Director Errol Drummond – who was one of the commissioners – argued for the scrapping of the dual labour system "as a matter of urgency" and for the legalisation of black trade unions so that they could participate in the collective bargaining process. At the federation's Annual General Meeting (AGM) that year, then SEIFSA President Dr JP Kearney supported Drummond's call, arguing that all forms of discrimination should be eliminated from labour relations. The following year, SEIFSA became a signatory to the Urban Foundation Code of Employment Practice and subsequently persuaded its labour partners that job discrimination and racially based provisions should be removed from the Main Agreement concluded in negotiations between the two parties.

Among the Wiehahn Commission's key recommendations were that full freedom of association be extended to all employees, that trade unions be allowed to be registered, regardless of their racial composition, and the phasing out of statutory job reservation. This led to two black trade unions being admitted as full members of the National Industrial Council for the Iron, Steel, Engineering and Metallurgical Industries in 1980. The following year, two more black trade unions joined the Council.

Then SEIFSA President Graham Boustred was subsequently to boast that the M&E sub-sector had been "the first major industry to eliminate completely the concept of race from our agreements and [that] the SEIFSA minimum wage has been a target of achievement for many other industries".

Although subsequent developments would not bear out the claim regarding the level of transformation in the sub-sector, as will be indicated in coming chapters, SEIFSA also claims to have steered its industry "to significant black advancement". As an example, the federation cites the fact that the wage gap between skilled and unskilled workers "had been narrowed from a ratio of 5:1 in 1961 to 2.8:1 in the early Eighties".

In 1988, the M&E sub-sector experienced its first national strike, which was organised by NUMSA and lasted for three weeks. Two years later, SEIFSA stated that, in partnership with the trade unions, it registered progress in helping black employees to buy houses. Collectively, they amended the rules of the Metal Industries Group Pension Fund to enable black workers to pledge their pension contributions as security for home loans.

Five years after the inaugural strike in the sub-sector, on 31 July 1992, NUMSA again embarked on a strike, this time in an environment that SEIFSA described as a period of tough economic conditions and escalating retrenchments. Involving about 80 000 NUMSA members, the strike lasted for a full month before an agreement could be reached with the federation. The tough economic conditions continued into the following year when employment in the sub-sector dropped to below 300 000.

# SEIFSA in the Democratic Era

In its self-curated history, SEIFSA states that, ahead of the founding democratic elections in 1994, it encouraged companies affiliated to its member associations to support the elections by calling on their employees to vote. When President Nelson Mandela's Government of National Unity subsequently established the Truth and Reconciliation Commission to investigate apartheid-era human rights violations in return for amnesty from prosecution for those who had been deemed to have told the truth, David Carson, Director of SEIFSA's Industrial Relations Division at the time, submitted a report to the commission. In it, he covered "events and developments" affecting the M&E sub-sector and SEIFSA from 1960 to 1994.

The dawn of democracy ushered in a new dispensation within the Metal Industries Benefit Funds Administrators (MIBFA), a Section 21 company which controlled the sub-sector's pension, provident, sick pay and permanent disability funds. From 1994 onwards, unions in the sub-sector enjoyed equal representation with employers on MIBFA's Board of Trustees.

As the new Government went about passing a series of laws intended to address the legacy of apartheid, Brian Angus, SEIFSA's Executive Director at the time, was concerned about what he considered to be an unwelcome interventionist trend in the economy. First in 1999 and again in 2003, Angus warned that the Government's interventionist trend would raise the cost of compliance, hamper economic growth and job creation, and make the economy less competitive. Angus's caution creates the mistaken impression that the M&E sub-sector (or, indeed, South Africa's manufacturing sector) had previously been internationally competitive. His claim was that the "outpouring of regulatory legislation across a broad spectrum of business activity" would have a deleterious effect on the economy. As has been indicated in the preceding chapters, South Africa's manufacturing sector had thrived during the apartheid era precisely because it was protected from international competition and enjoyed considerable government support.

SEIFSA prematurely claims that, in the new democratic dispensation, it – on behalf of its members – and unions active in the sub-sector concluded a "New Deal" in industrial relations, which saw the confrontational style of collective bargaining replaced with "greater mutual understanding and a more conciliatory bargaining approach". As part of this "New Deal," SEIFSA claimed, the focus during negotiations on wages and conditions of employment was on the rapid resolution of differences and "the achievement of longer-term gains across a broader range of issues".

Since its inception, SEIFSA was funded by its member associations. Each company affiliated to an employer association which was a member of SEIFSA paid membership fees to that association and to SEIFSA, for the benefits of being part of a powerful employers' federation. In addition, these companies paid a levy to the federation for each person that it employed who was covered by the bargaining council. That model worked well when SEIFSA had many associations as members, and when the companies belonging to the associations collectively had many employees on their books. During tough economic periods, as companies laid employees off, SEIFSA's income was negatively affected.

Angus, who was appointed Executive Director in 1987 and remained in office until 2009, says SEIFSA's finances were in a parlous state when he joined the federation because companies were facing difficulties at the time. To supplement SEIFSA's membership fees, Angus came up with a brilliant innovation of introducing a number of training and consulting services in industrial relations, economic and commercial fields, skills development and artisan training, for which the federation charged its members a fee (Van Biljon, 2014).

"It became clear to me early on that we were on a one-way street to nowhere. Every year the number of employees in the industry was getting less and less, yet we could only increase our membership levies on a per-capita basis, and at about the rate of inflation. So, our income in real terms was diminishing every year. The only way out was to become more professional and to charge for some services. Members responded positively, and that not only helped SEIFSA survive difficult years in the Nineties, but [it] strengthened us as an organisation," Angus explained.

These services proved to be popular with member companies, the majority of which are small and medium-sized, hence they do not have all this expertise internally. In addition, SEIFSA "severely" reduced its administration costs, including personnel in that area of the business. As Angus was to say later in a SEIFSA supplement published in the *Financial Mail*, this innovation "worked very well and remains a key strategy of the federation" (Van Biljon, 2014).

In addition, SEIFSA and the trade unions in the sub-sector took advantage of the provision in the Labour Relations Act of 1995, which provides for the entities which represent the majority of employers and employees, respectively, in an industry to charge a levy on those within the same industry who are members of neither entity but benefit from the negotiated settlement or the stability resulting from it. This is called the collective bargaining levy (CBL). Following the successful conclusion of wage negotiations in 2003, SEIFSA and the labour unions lobbied the Minister of Labour to gazette the CBL for all employers and blue-collar workers in the M&E sub-sector, with effect from May 2003. The CBL was valid for five years and was renewed for the same period in 2008. Between 2003 and 2012, SEIFSA received R52 666 790 from the CBL.

Angus argued that the industry-wide CBL was "fully justified". He explained that, until then, it was only companies that were members of SEIFSA-affiliated employer associations which bore the full cost of negotiating agreements which benefitted "all the 8 000-plus employers in the industry". He felt that "the very modest levy" of R150 per month per company, payable to SEIFSA, would ensure that the negotiation cost burden was spread more equitably among all the companies which benefitted from the negotiated settlement, called the Main Agreement.

The CBL, Angus revealed, had the effect of persuading non-affiliated companies to take up membership of employer associations which were members of SEIFSA. This was because they wanted to enjoy more benefits than just those which flowed from the negotiated settlements.

When the second CBL expired on 31 December 2012, it was not renewed because a relatively new player on the Metal and Engineering Industries Bargaining Council (MEIBC), called the National Employers Association of South Africa (NEASA), lodged a dispute challenging SEIFSA's status as the majority employers' representative on the Council. According to then SEIFSA President Henk Duys, the CBL's non-renewal "had serious negative funds flow consequences" for SEIFSA and NUMSA (SEIFSA Annual Review, 2013).

NEASA challenged the allocation of seats among employers' associations on the MEIBC. Until then, SEIFSA, as a federation, had represented its member associations on the council. However, NEASA argued that SEIFSA-affiliated associations were "nothing more than empty shells" and contended that, since it was not itself a registered employers' organisation, SEIFSA had no right to participate in the MEIBC as an agent of its member associations. This precipitated a major crisis for SEIFSA, which for decades had been the only employers' representative on the bargaining council, and it led to "protracted arbitration proceedings" between SEIFSA-affiliated associations and NEASA (SEIFSA Annual Review, 2013).

The truth, however, as Duys reported at SEIFSA's 2013 AGM, was that SEIFSA member associations had more employers in the sub-sector that, collectively, "employed three times more employees" than did all the other independent employer associations represented in the MEIBC. Duys expressed confidence about the outcome of the aforementioned legal arbitration proceedings, after which, he felt, the MEIBC's constitution would have to be reviewed and amended to "more equitably reflect the current economic state" of the M&E sub-sector in the country (SEIFSA Annual Review 2013).

Although SEIFSA was responsible for central collective bargaining, with labour, on behalf of its members, Angus nevertheless subsequently held the view that over time the M&E

sub-sector must consider transitioning to shop-floor bargaining. He argued that Britain had moved away from collective bargaining, that moves were afoot to do the same in Australia and that Germany – which was "the home of centralised collective bargaining" – was having a re-think on the issue. The new trend, he said, was to have most matters dealt with at plant level, and fewer matters – such as agreements on provident and pension funds as well as training and development programmes – handled centrally. He was also opposed to the extension of collective bargaining agreements to the entire sub-sector, arguing that this would lead to non-affiliated companies cancelling their plans to expand and retrenching employees (Angus, 2022).

However, Angus's musings go against the very reason for SEIFSA's formation way back in 1943. They also ignore the fact that the majority of companies which are members of SEIFSA, through their employer associations, employ no more than 50 people, hence they do not have the time or capacity to handle negotiations internally. There is also the matter of the little dynamic that internally-held negotiations have the potential to strain relations in a company between employee representatives and those in charge.

Following the implementation of the Sector Education and Training Authority system in 2000, in terms of the Skills Development Act of 1998 and the Skills Development Levies Act of 1999, SEIFSA also played an important role in the establishment and the management of the Manufacturing, Engineering and Related Services Sector Education and Training Authority (MerSeta), and the introduction of the new training levy system. The federation also continued to be active in the training of artisans. In 2003, it outsourced to GijimaAST the management of its apprentice training centre in Benoni and renamed it the Fundi Training Centre.

In the same year, the federation also became a founding member of Business Unity South Africa (BUSA). Raymond Parsons, former Director-General of the South African Chamber of Business, described SEIFSA's role in the apex business organisation as follows: "SEIFSA has been a reliable resource centre from a key sector of the economy for leadership, professionalism and expertise in organised business policy formulation, but without wishing to dominate the situation. It has been sensitive to finding the balance between minding its own business and participating in the bigger picture" (Van Biljon, 2014:7).

SEIFSA also piloted an innovative artisan training programme, which was subsequently adopted and funded by MerSeta. In response to growing demand, SEIFSA funded a major expansion of the Fundi Training Centre, using the reserves of the previous Metal and Engineering Industries Education and Training Fund.

When he retired in 2009, Angus was succeeded by David Carson, SEIFSA's former Director of Operations. Unlike Angus's, Carson's tenure as Executive Director was relatively short. He stepped down a mere four years later, in April 2013, after a 37-year career with the federation. Deputy Chief Executive Officer Elsa Venter was appointed Acting Executive Director with effect from May 2013.

In his report as SEIFSA President and Board Chairman in the annual report of that year, Henk Duys bemoaned the fact that senior industry leaders who should be participating in the affairs of the federation were "not coming forward and volunteering their services and input". He observed that in the first 55 to 60 years of its existence, SEIFSA was led by "captains of industry – the Chairmen and Chief Executive Officers of large corporations which were instrumental in building our country." He said that it was expected that, as part of their career paths, these captains of industry would, in addition to their positions in their respective companies, go through SEIFSA's ranks and serve at least a term as its President (SEIFSA Annual Review, 2013).

At the time, SEIFSA had 27 employer associations as members, which had a combined membership of 2 192 companies which collectively employed 221 942 workers covered by the industry's Main Agreement (SEIFSA Annual Review, 2013).

In his personal observations in his address to the 2013 AGM, Duys commented on the difficult state of the economy and the country's poor governance. He said that, in its 70-year history, "there cannot be too many periods when [SEIFSA] has been faced with so many challenges – all at the same time". He said it was no exaggeration that the turbulence that the organisation was experiencing was "probably the most challenging in our history".

Turning to the forthcoming wage negotiations with labour in 2014, Duys said many of SEIFSA's smaller member associations insisted on "special dispensations" in those talks. Motivating his point, he added:

> Changed circumstances means [sic] changed approaches to the old problems are needed. We will split our organisation if large business dictates to small business, and yet each has legitimate cases for their points of view. The pressures for more regional or sector autonomy, for greater flexibility and freedom of action at company level, for wage and cost relief to allow business survival – not just growth – are growing by the day. All this speaks to how we bargain, what bargaining model should be adopted – fully central as in the past, partial[ly] central and partial[ly] regional / sector or fully regional / sector. Centralised bargaining with special dispensations for motivated regional / sector needs would seem to be the answer, but individual associations need to make up their own minds and develop appropriate mandates for the negotiating teams to implement.

Duys called for a change in SEIFSA's approach to negotiations, arguing that the way the federation bargained in the past "was self-destructive". Companies affiliated to associations federated to SEIFSA did not "have the money to satisfy the much-needed and justified demands of our workers", he said. Instead, both the companies and labour needed to commit jointly to sacrificing "short-term gains in favour of jointly tackling productivity, enhanced global competitiveness and expansion of our economy in the medium to long terms". He denounced one-sided bargaining, which presumably had characterised SEIFSA until then, emphasising that member companies were seeking "real benefits".

Then Duys introduced a wholly new concept: before even beginning with the negotiations the following year, SEIFSA must raise with the labour leadership "how we are going to deal with the violence and intimidation question". This was to form the central plank of SEIFSA's approach to the 2014 wage negotiations.

## Conclusion

SEIFSA's fortunes have mirrored those of the metals and engineering sub-sector, whose fortunes has been consistent with that of the broader manufacturing sector to which it belongs. It has been clear in this chapter that, while it was formed to represent the interests of M&E employers as a collective bargaining agency, over time the federation had to extend its scope to offering paid-for consulting and training services in areas such as industrial relations, economic and commercial, skills development, and health and safety in order to supplement its income.

Former SEIFSA Executive Director Brian Angus (1987–2009) has made it plain that, as the economy took a turn for the worse, more companies in the sub-sector retrenched employees. Given that SEIFSA's membership model is based, in part, on the number of blue-collar workers employed by each of the companies affiliated to an association which is a member of the federation, SEIFSA's finances were negatively affected. Consequently, Angus and his team had to find ways in which to supplement the federation's membership fees to ensure its continued sustainability. That included a "severe" reduction of the federation's administration costs and staff.

In the next chapters, we see how the challenge confronted by Angus during his tenure continued to worsen.

# Part II

# Chapter 4

# The Beginning of a New Era

## Introduction

In October 2013, I was appointed Executive Director of SEIFSA, the first – and so far, at the time of writing, only – black person to hold that position in the organisation's 80-year history. My appointment came into effect on 1 November 2013.

I had responded to an advertisement in the *Sunday Times* that called for applications for the position of Executive Director. Until then, I had not heard of SEIFSA, despite my 15-year career in the media (including as a newspaper Editor) and my subsequent 10-year career as a senior business executive. It was when I read up in preparation for the interview that I got a sense of what the organisation did and what it stood for. However, I knew more than enough, for instance, about the Chamber of Mines, which did the same thing for employers in the mining sector that SEIFSA did for employers in the M&E sub-sector of manufacturing.

The shortlisting was handled by a recruitment agency, which received all the applications, and I was interviewed by three men – President Henk Duys, Vice-President Norbert Claussen, co-opted Board member Neil Penson – and SEIFSA Deputy Chief Executive Officer (CEO) Elsa Venter, who had been Acting Executive Director since May that year. Duys was Executive Chairman of his own company, Durban-based Duys Engineering Group, Claussen was Managing Director of his own company, Cape Town-based S&G Signs (Pty) Ltd, and Penson was Company Secretary of Johannesburg-based Babcock International Group. I was, therefore, interviewed by an all-white team made up of three men and one woman.

In the course of the interview, Duys informed me that the initial intention had been to fill the job with an internal candidate. To this end, the vacancy had been advertised within SEIFSA and among the federation's 27 member associations. He also told me that there was an internal applicant for the position, but the panel did not consider him ready for appointment. It was only when a suitable internal candidate could not be found that the job was advertised externally.

Despite its racial composition, the panel made it clear that it felt that it was time that SEIFSA was transformed, hence it was looking for somebody who would be able to transform the organisation. Second, they were concerned about the organisation's relatively low profile outside the M&E sub-sector and wanted somebody who would be able to raise its public profile and allow it to be taken seriously by policy makers, among others. It was during the interview that I learned about another organisation, NEASA, which was a major bugbear for SEIFSA. I was told about it and its nuisance value, but I was not specifically asked if I would be able to counter its virulent anti-SEIFSA propaganda.

At the reception, I was met by a white receptionist, who led me to the waiting room nearby. As I sat in there, with the door open, I could see people passing by in either direction in the passage, and none of them was black. I subsequently understood why transformation was deemed one of the priorities for whoever would emerge as the successful candidate. In that case, I guess the fact that I had successfully led a transformation project elsewhere, when I relaunched *The Saturday Paper*, which catered for a white and Indian readership, as *The Independent on Saturday*, which catered for all races in KwaZulu-Natal, had worked in favour of my candidature.

I had already accepted a job at the University of KwaZulu-Natal when, a few weeks later, I finally received the appointment letter from SEIFSA. I was due to start at the university on 1 November 2013, so I had to turn it down to accept the SEIFSA appointment once the federation had responded favourably to my request for some amendments to its initial offer. Since it was still October, I had some weeks during which to prepare for the move.

The following image shows the composition of how the SEIFSA "Executive Committee" looked that year:

## SEIFSA
## **Executive** Committee

> **President**
> HW Duys

> **Vice-Presidents**
> M Pimstein | N Claussen | B Ashlin

> **Additional Members**
> C Davis | M Garcia

> **Co-opted**
> C Dawtrey | DCG Murray
> M Dames | N Penson

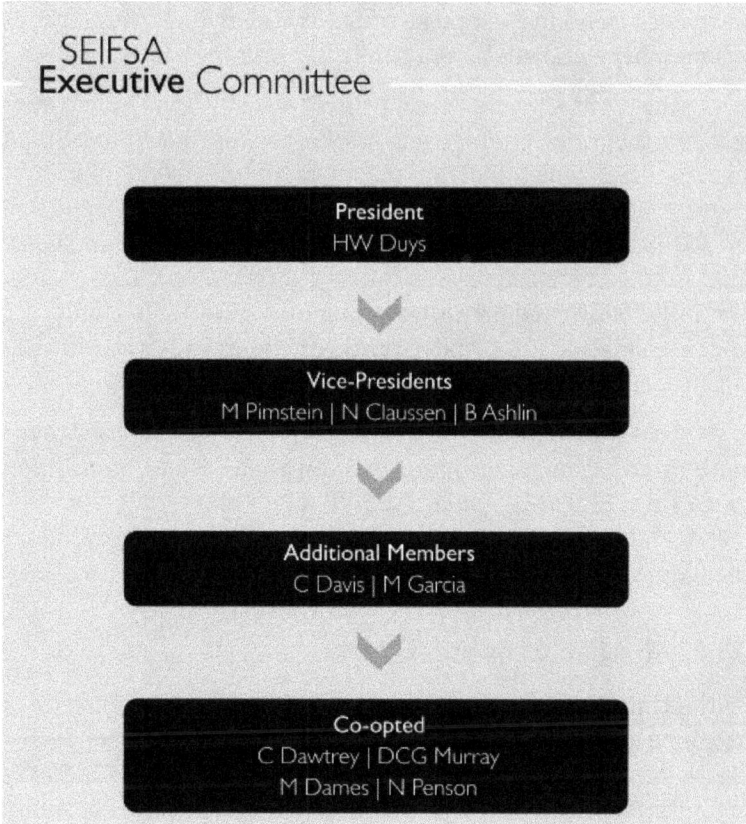

*Source: SEIFSA Annual Report 2013*

At the time, SEIFSA was in the process of reviewing its strategy, and Venter had arranged some strategy sessions. She invited me to some of the sessions. The one on 24 October 2013, a week before I was due to start with the federation, was particularly insightful. I was even invited to the federation's end-of-month Staff Communication Session. It was during the strategy session on 24 October 2013 that I learned that all was not well with the organisation. SEIFSA was not in good health at all. I learned, for instance, that the CBL, which had expired on 31 December 2012, had contributed about R6 million annually to SEIFSA's coffers and that, without it, SEIFSA had budgeted for a loss of R8.1 million for the 2013/14 financial year. I learned further that there was a downward trend in the uptake of SEIFSA's products and services by "member companies". Only in-house training was up.

The phrase "member companies" refers to those companies that are members of employer associations affiliated to SEIFSA and are themselves indirect members of the federation. "In-house training", on the other hand, refers to training conducted by SEIFSA employees on the premises of these companies; this is distinguished from training offered on the SEIFSA premises, which requires that companies send their employees to the federation's 42 Anderson Street, Marshalltown office in downtown Johannesburg.

That was all news to me. None of it had been disclosed to me by the panel during my interview in August. Clearly, in addition to raising SEIFSA's public profile and countering NEASA's propaganda efforts, I would have to find a solution to these "new" challenges of which I had just become aware.

## Worrying Early Discoveries

By the time I assumed my role as SEIFSA's Executive Director, I had done even more reading on the federation and had conducted a desk-top benchmarking exercise comparing it to other industry organisations, such as the Chamber of Mines, the Consumer Goods Council of South Africa, Business Leadership South Africa, Business Unity South Africa, Business Against Crime, and the National Business Initiative. Both in terms of Vision and Mission and in terms of the corporate governance structures, SEIFSA was very different from these organisations. It was clear to me that those were among the first anomalies that we had to tackle as part of the strategy review process.

I was subsequently to discover more issues that required urgent attention, such as a very deep-seated culture of nepotism and indolence at SEIFSA. Based on my prior experience and my Master of Business Administration studies, I knew that it would not be easy to bring about the kind of radical change that was necessary, but I was determined to ensure that the job was done. I was clear in my mind that my legacy at SEIFSA was going to be a more results-focused culture and a sound corporate-governance structure that would be appropriate for a modern, business-oriented organisation.

I was also aware of the amount of anxiety that existed among SEIFSA employees about the organisation's future under its first black leader. Most of them had worked for the federation for many years (some for two or even three decades) and had grown comfortable in an untransformed environment that felt like their second home. It was important, therefore, both for staff members and myself, that I introduce myself – as a person and a leader – from day one and clearly spell out my expectations. In my experience, it is very important to ensure that everybody has a shared understanding of the future and good appreciation of the expectations that one has of them. With this in mind, I prepared my first presentation at my leisure at home and asked Venter to convene a meeting with the SEIFSA Executives and Managers for 9am on my first day.

The first of November 2013 was a Friday, and two important meetings were scheduled for Monday, 4 November 2013: a SEIFSA "Executive Committee Meeting" in the morning and a SEIFSA Council Meeting that afternoon. "Executive Committee" was the odd nomenclature used at SEIFSA to describe what was essentially a Board of Directors, while the SEIFSA Council was what I would later describe as an assembly of member associations, where each was represented by two people. Also included in the SEIFSA Council were several co-opted members. Where historically such co-opted individuals were captains of industry who would add gravitas and expertise to the body's deliberations, in later years these were Human Resources or Industrial Relations managers who were neither Chairmen nor Secretaries of their affiliated associations.

At the 9am meeting on 1 November 2013, I made a presentation to SEIFSA Executives and Managers. Headed "Building on A Proud Legacy", the presentation covered the points discussed below. I thanked everybody for having welcomed me warmly, expressed my appreciation to Venter for having invited me to the strategy sessions before I started in the job, commended her on her "reliable stewardship of SEIFSA over the past five months", and stated that I was aware of SEIFSA's proud, 70-year history and was "very conscious of the fact that I stand on the shoulders of giants who built the organisation to be what it is today". I pledged that, working together, we would be "more than equal to the challenges of our time and take SEIFSA from strength to strength".

Next, I spelt out my vision for SEIFSA in five bullet points:

- "SEIFSA must *Regain its Voice* as a matter of extreme urgency;
- "We must have a SEIFSA that is indubitably recognised as *The Authoritative Voice* of the steel and engineering industries;
- "We must *improve SEIFSA's Public Profile*;
- "SEIFSA must be seen to be offering *Real Value for Money* to its members; and
- "SEIFSA must be seen to have *South Africa's Best Interests at Heart.*"

I knew that I was fortunate to have joined SEIFSA at a time when it was reviewing its three-year strategy. That offered me the opportunity to be co-architect of that strategy and to take ownership of it. So, in that presentation, on a slide headed "What do we need to do?" I informed the Executives and Managers that, as we worked on the new three-year strategy, we needed to "ask ourselves frankly what SEIFSA's value proposition is," and "interrogate what the SEIFSA brand stands for now and what it should stand for over the next five to ten years" – and that in the process we would have to be prepared "to engage openly, robustly and yet constructively, with nothing deemed to be sacrosanct".

Next, I said that there were questions that we needed to answer honestly. These were whether SEIFSA's current vision ("Providing Industry Solutions") was sufficiently aspirational and did justice to the federation, whether the organisation was structured appropriately and whether it leveraged its collective expertise appropriately or functioned more as divisions rather than an organisation. I also observed that SEIFSA had no corporate values, but only some loose-sounding statements listed as "staff values". I stressed the need for us to develop such corporate values, values we would all need to embrace and live. I also emphasised the introduction of on-going risk management, something that was sorely lacking at SEIFSA. For instance, I had asked for – but not received – the organisation's Risk Register.

I informed the Executives and Managers that we would have three months within which to discuss all these matters so that I would be able to present recommendations to the "Executive Committee" at its next meeting in February 2014.

Finally, I introduced myself as a person and a leader. On a slide headed "Finally, who is Kaizer?" I stated the following:

· "I am a man of integrity – and I expect the same from those with whom I work;
· "I work very hard and lead by example;
· "I am open-minded and encourage discussion;
· "I have a strong sense of fairness; and
· "I am decisive, after consideration of all relevant facts".

I concluded the presentation by expressing confidence that, working together as a team, we could take SEIFSA "to new heights", and I invited everyone to work hard to make it possible for us to achieve that goal. I told everybody that I looked forward to working with them.

My second day on the job was spent at the Country Club Johannesburg in Auckland Park, where SEIFSA "Executive Committee" and Council Meetings were held. Both meetings were quite a revelation. In addition to the elected "Exco Members", in attendance were all the SEIFSA Executives, who each gave updates on their respective areas of responsibility (Industrial Relations, Skills Development, Economic and Commercial, Health and Safety, etc.). The only decision to be made by both structures that day was that I had to see to the resumption of the legal arbitration process with NEASA, which had been placed in temporary abeyance; all member associations committed to getting their member companies to contribute towards a legal fund to cover the cost of arbitration.

However, I did get to learn one important thing at my inaugural meeting of the SEIFSA "Executive Committee" – and that was that Scaw Metals Chairman Ufikile Khumalo, who was also a senior executive at the Industrial Development Corporation, was endorsed at the October AGM to be SEIFSA's next President. Although he was not present at the meeting, it was reported that he was likely to be at the next one in February 2014. For the very first time in its 70-year history, SEIFSA was to have a black President and a black Executive Director!

According to the organisation's Memorandum of Incorporation, former Macsteel Chief Executive Officer Michael Pimstein, who was the First Vice-President at the 2012 AGM, was to have succeeded Duys as President. The First Vice-President automatically succeeded the President, who served a year in office, and the out-going President automatically became Second Vice-President. That meant that Pimstein was due to be President again, with Duys as Second Vice-President, with a new First Vice-President elected at the 2013 AGM. However, Pimstein was no longer available to assume the SEIFSA presidency, so the October 2013 AGM had approved the election of Khumalo as President. Aveng Trident Steel Managing Director Alpheus Ngapo was elected First Vice-President.

On average, the "SEIFSA Executive Committee" meetings took between two and three hours and the SEIFSA Council meeting took about two hours. The first meeting occurred quarterly, but the latter was monthly. In addition, there was also a Chairman's Forum, made up of chairpersons of the member associations, which met monthly at SEIFSA's offices. It was a major duplication because the self-same chairpersons of associations were also at the monthly SEIFSA Council meetings.

Although SEIFSA claimed in the previous chapter to have steered the M&E sub-sector "to significant black advancement", both the federation itself and most of its member associations and affiliated companies were anything but transformed. To start with, SEIFSA itself was not representative of South Africa *circa* 2013, and the SEIFSA council was a sea of white faces, with a sprinkling of black faces holding relatively junior positions in their respective companies. Among the 27 member associations, only one had a black chairperson (Edward van Ryneveld of the Light Engineering Industries Association of South Africa), and the only affiliated private-sector company that I knew to have a black

Managing Director was Austrian-owned voestAlpine VAE SA (Pty) Ltd, whose Managing Director was David Marite (it is now headed by Pulane Msimang-Kingston).

In a staff complement of more than 30, not only were black people (including coloured and Indian people) in a minority, but they also held the most junior positions. As Skills Development Executive, Nazrene Mannie was the only black Executive and, as Health and Safety Manager, Zimbabwean Nonhlalo Mphofu was the only black Manager. Among the staff members, more than two-thirds did not have a tertiary qualification. Instead, they had been with the federation for many years and had advanced from one position to the next over the years. Some had worked for SEIFSA throughout their lives and knew no other employer and no other corporate culture.

The photograph below depicts the profile of the SEIFSA staff in September 2013, as captured in the organisation's 2013 Annual Report:

*Source: SEIFSA Annual Report 2013*

To top it all, the place was teeming with nepotism. Former Executive Director Brian Angus's son, Gordon, was employed as Industrial Relations (IR) Executive, without any post-matric qualification, and Brett Carson, the son of Angus's successor, was an IR Officer. Angus led a team made up of graduates (Brett

Carson, IR Manager Michael Lavender, who had been in SEIFSA's employ for two decades, and a young black intern who was a graduate of the Tshwane University of Technology). Operations Director Lucio Trentini's wife was a service provider to SEIFSA and provided meals and other edibles for the monthly Staff Communication Meetings and training sessions hosted by the organisation, and some staff members in the west wing of the building were rumoured to be related to Venter.

Structurally, SEIFSA was divided into two wings, east and west, and these were the dominions of Trentini and Venter, respectively. Not only were their respective offices situated in those wings, but staff members were divided neatly into those who fell under Venter and those who fell under Trentini and had their offices in the appropriate wing of the sixth floor of Metal Industries House. Biometric scanning machines were placed at the doors leading to the entrances of the two wings, and everybody was required to have their fingers scanned on the machines as they entered or left those wings. Although they were all on the same floor of the building, there were two very clear parts to that floor, with those who belonged to one wing being strangers to the other wing.

There was a proprietorial attitude towards each wing on the part of Venter and Trentini and a very clear sense of belonging to one or the other wing among staff members. Some employees would tell me openly: "We don't go to that side of the building" – unless duty required them to do so. I found the situation absolutely intolerable. At meetings, whenever colleagues from the east wing raised concerns about debtor management or data integrity, there was great defensiveness on the part of their colleagues from the west wing, with Venter at the forefront of the defensiveness.

In that structure, only Venter and Trentini reported directly to the Executive Director. The Executive Director, it seemed, was tolerated in the first office in Trentini's side, the east wing, with Trentini occupying a corner office closest to the bathroom on the right-hand side.

According to their LinkedIn profiles, Venter had joined SEIFSA in July 1983 and Trentini had done so in September 1990.

The following functions reported to Venter and were housed in the west wing: Association Administration, Finance, Marketing and Communications, as well as Information Technology. Housed in the east wing were the following functions, with people reporting to Trentini: IR, Economic and Commercial (EC), Health and Safety, as well as Skills Development. The structure, as captured in the 2013 SEIFSA Annual Report, is reproduced in the following image:

*Source: SEIFSA Annual Report 2013*

No organisational structure is published in the 2011 or 2012 SEIFSA Annual Reports, which are titled "Report of the Executive Director"; hence, I do not know whether the structure had always been like that or if it was how Venter and Trentini had decided to divide up responsibilities between themselves following David Carson's departure. As it stood, the structure appeared to have been designed to ensure that the new Executive Director would be no more than a sinecure. It had so entrenched Venter's and Trentini's powers in the federation and considerably weakened the in-coming Executive Director, who was meant to manage SEIFSA indirectly through them.

I could not fathom the logic of either the structure or the designations. I could not understand why an organisation the size of SEIFSA needed a Deputy CEO, nor could I make sense of the Operations Director designation.

A strange feature of the federation's annual reports was that, until then, they hid more information than they revealed. Instead of being annual reports that reported to the membership and other stakeholders about SEIFSA's performance during the year under review, they merely profiled the federation, giving its history, listing its member associations and the products and services it provided, and reporting on what the respective divisions had been doing during the year. The only reports on the year under review were provided by the president and the Executive Director respectively. So opaque were the reports that there was absolutely no mention of the federation's financial performance during the years in question, nor were there any other important disclosures.

The walls on the passage leading to the east wing featured the proud history of the federation, replete with pictures of prominent individuals in that history during the era in question. Similarly, the meeting rooms were named after prominent figures in SEIFSA's history. Among those I still recall are the Drummond Room and the Angus Room (interestingly, there was no Carson Room). There was also a big room with old, rickety tables and desks, which was very sparingly used, on the walls of which the photographs and names of past SEIFSA Presidents were prominently displayed. It may not have been the intention, but everything about the place said to me that I was a trespassing alien in that space. As a black person, I did not feel welcome there.

Some four years into my tenure, I resolved to have the various meeting rooms renamed so that none of us would continue to feel alienated. I thought that having staff members rename the rooms would ensure that they would have a greater sense of belonging and co-ownership. After having gone through the suggested names, I selected Rainbow Room for the old Drummond Room, Indaba Room for the old Angus Room, and Dinaledi (a Sesotho word that means "stars" in English) and Sycamore for the other two rooms whose original names I have long forgotten.

Since the big room with old, rickety tables and desks was hardly used, I moved swiftly to have it turned into a beautiful, modern Boardroom with a U-shaped table, and we used it for our

Executive Committee meetings. Sometimes, the room even hosted Board meetings on occasions when there were no SEIFSA Council Meetings to follow. The Boardroom was big enough for Board Meetings, but it could not accommodate the 60-plus people who normally attended meetings of the SEIFSA Council. The pictures that used to have pride of place in that room were safely kept in storage.

On the walls of that Boardroom, we continued to update the details of the federation's presidents on a brass plaque. The list of Presidents, from the organisation's founding until 2013, appears in the following image:

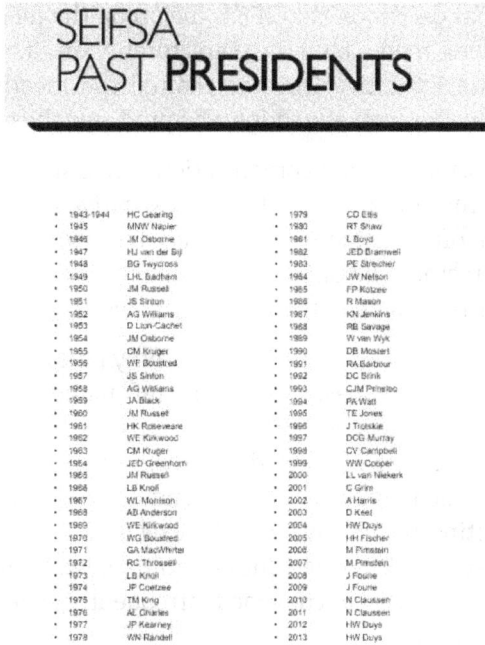

## SEIFSA PAST **PRESIDENTS**

| Year | Name | | Year | Name |
|------|------|---|------|------|
| 1943-1944 | HC Gearing | | 1979 | CD Ellis |
| 1945 | MNW Napier | | 1980 | RT Shaw |
| 1946 | JM Osborne | | 1981 | L Boyd |
| 1947 | HJ van der Bijl | | 1982 | JED Bramwell |
| 1948 | BG Twycross | | 1983 | PE Strecher |
| 1949 | LHL Badham | | 1984 | JW Nelson |
| 1950 | JM Russell | | 1985 | FP Kotzee |
| 1951 | JS Sinton | | 1986 | R Mason |
| 1952 | AG Williams | | 1987 | KN Jenkins |
| 1953 | D Lion-Cachet | | 1988 | RB Savage |
| 1954 | JM Osborne | | 1989 | W van Wyk |
| 1955 | CM Kruger | | 1990 | DB Mostert |
| 1956 | WF Boustred | | 1991 | RA Barbour |
| 1957 | JS Sinton | | 1992 | DC Brink |
| 1958 | AG Williams | | 1993 | CJM Prinsloo |
| 1959 | JA Black | | 1994 | PA Watt |
| 1960 | JM Russell | | 1995 | TE Jones |
| 1961 | HK Roseveare | | 1996 | J Trotskie |
| 1962 | WE Kirkwood | | 1997 | DCB Murray |
| 1963 | CM Kruger | | 1998 | CV Campbell |
| 1964 | JED Greenhorn | | 1999 | WW Cooper |
| 1965 | JM Russell | | 2000 | LL van Niekerk |
| 1966 | LB Knoll | | 2001 | C Grim |
| 1967 | WL Morrison | | 2002 | A Harris |
| 1968 | AB Anderson | | 2003 | D Keet |
| 1969 | WE Kirkwood | | 2004 | HW Duys |
| 1970 | WG Boustred | | 2005 | HH Fischer |
| 1971 | GA MacWhirter | | 2006 | M Pimstein |
| 1972 | RC Throssell | | 2007 | M Pimstein |
| 1973 | LB Knoll | | 2008 | J Fourie |
| 1974 | JP Coetzee | | 2009 | J Fourie |
| 1975 | TM King | | 2010 | N Claussen |
| 1976 | AE Charles | | 2011 | N Claussen |
| 1977 | JP Kearney | | 2012 | HW Duys |
| 1978 | WN Randell | | 2013 | HW Duys |

*Source: SEIFSA Annual Report 2013*

Nowhere else had I come across something similar or felt the way that SEIFSA made me feel as a black executive. Clearly, there was a lot of work ahead for me. I knew that I was going to upset people with vested interests by tackling these challenges but, buoyed by the mandate given to me by the interviewing panel to transform the organisation, I was more than prepared to do so.

The next surprise that awaited me was that a fortnightly meeting of "the Directorate" was scheduled to take place in my office. I accepted the invitation, even though I did not know what "the Directorate" was. A big book in which minutes of the meeting were kept was stored in my office, and the minutes of the last meeting were duly shared with me. That last meeting had been between Venter and Trentini, who jointly constituted "the Directorate" at the time. I was soon to learn that the label referred to the three Executive Directors of SEIFSA, namely the so-called Executive Director, the Deputy CEO and the Operations Director. All operational decisions at SEIFSA, including the most mundane or basic, were made by "the Directorate" in its fortnightly meetings. The Executive Director chaired the meeting and the Operations Director was the scribe who produced the minutes.

It was strange that an organisation with a staff complement of only just more than 30 had three Executive Directors, when large – and often listed – companies tend to have only two Executive Directors, the CEO and the Chief Financial Officer (CFO). Be that as it may, that was the situation which obtained at SEIFSA.

Among the things on the agenda of my first meeting of "the Directorate" was a request by a junior staff member for a loan of about R12 000. That was what the three most senior people within the federation had to concern themselves with! The other items were of the same magnitude. When I enquired whether there was another meeting which involved the executives, I was told that they were part of a Management Committee, which included the Managers, and which was convened infrequently, whenever it was deemed appropriate.

As I recall, I could tolerate only one more meeting of "the Directorate" before I abolished the structure and replaced it with Executive Committee meetings that included the three of us and all the Executives.

Another big concern for me was the presence of a smoking room, prominently labelled as such, just next to the waiting area, with its door opening onto the passage leading to the east wing of the building. Staff members and those who were attending training at SEIFSA were encouraged to smoke there, and when the door was open the smoke wafted into the passage and affected us all. I was most unhappy about this arrangement since I knew that, during her time as Minister of Health during the early days of our democracy, Dr Nkosazana Dlamini-Zuma had outlawed indoor smoking. I could not understand how a socially responsible corporate citizen like SEIFSA continued to tolerate a patently illegal practice.

Interestingly, this happened even though the landlord, MIBFA, had a firm rule that its employees who smoked should do so only in the open-air area set aside for smokers on the top of the seventh floor of the building. That meant that while the landlord was legally compliant as far as the country's smoking laws were concerned, SEIFSA as a tenant was not.

Within a matter of days after I had started in the job, I discovered that employees were meant to physically sign themselves in and out when they arrived and when they left the building. The receptionist was the custodian of the book where employees signed in and out. I was appalled that a professional services organisation like SEIFSA required its employees, part of whose job was to travel to member companies for training or consultation and to attend meetings of fraternal organisations, to "clock in" as though they were factory workers. When I asked why the practice was found necessary, I was told that some employees were guilty of spending a lot of time away from their workstations on the pretext that they were working.

## Changes Introduced Swiftly

I belong to the school that holds that any contemplated change should be introduced within the first 90 days. Not only does that emphatically signal one's intentions, but it also helps those who have been with the company for a long time to understand that things will not be the same again and to make peace with it. Similarly, it affords those who may be vehemently opposed to the envisaged changes an opportunity to review their own positions and to re-evaluate their futures.

Therefore, I moved swiftly to address the challenges mentioned in the previous section. I took the time to explain the rationale for my actions in an effort to take as many people as possible along with me on the journey to the new SEIFSA. In my first month in the job, I started with those things that were in my direct control to achieve some quick wins.

Firstly, I convened a staff meeting at which I announced that, going forward, any forms of nepotism would not be tolerated at SEIFSA. While nothing would be done to those already in the federation's employment, thenceforth nobody who was related to anybody working at SEIFSA would be eligible for employment at the federation, and family members would no longer be allowed to do business with the organisation. I announced that we would soon introduce a policy on conflict of interest (there were no policies at SEIFSA, except for a very few basic ones on human resources) and that all of us would be required to declare our business interests in any entities.

The news about Trentini's wife being the official caterer at SEIFSA's events and functions was shared in confidence with me, by one of the Executives, after that meeting. I called the Trentini into my office and asked him if what I had been told was true, and he confirmed that it was. He told me that my predecessor knew about the arrangement and had no problem with it. I then informed him that SEIFSA's business relations with his wife would be terminated immediately, and I subsequently informed the relevant colleagues in the Marketing and Communications and Finance Divisions of this so that they could act on that information.

In the same meeting, I also abolished the clocking system. I informed everybody that we needed to introduce a new culture of business responsibility at SEIFSA, one which would see everybody held accountable for their conduct. While we all had to ensure that we contributed meaningfully to the federation, in accordance with our respective job descriptions, we also had to earn our salaries and hold one another responsible. So, those in positions of authority had to know the whereabouts of those who reported to them and to hold them accountable for unexplained disappearances from work. That is, the default had to change from assuming that all employees were dishonest to regarding them as reliable and honest individuals, while holding accountable those given to truancy.

Also, within the first few weeks in the job, I changed the reporting structure and the east wing–west wing dynamic. Rather than having just the Operations Director and the Deputy CEO reporting to me as Executive Director, I also assumed direct responsibility for the following functions: Finance, Marketing and Communications, and EC. That saw the Finance Executive (as the position was called at the time), the Marketing and Communications Executive and the Chief Economist also reporting to me. I also changed the Skills Development Executive's reporting line from the Operations Director to the Deputy CEO. That left Venter with Skills Development, New Business Development, Association Management and Information Technology as the functions reporting to her, while Trentini held onto IR (his undoubted area of expertise) and Health and Safety.

The abolition of "the Directorate" and its replacement with a functional Exco took effect from January 2014, and it met monthly and jointly deliberated on all matters affecting the federation. This was revolutionary change at SEIFSA. Consequently, it took a while before each of the Executives participated enthusiastically in Exco meetings. The Directors found it difficult to accept that their word suddenly counted for no more than that of an Executive, and that, instead of relying on *ipse dixit*, they needed to persuade their colleagues to a particular point of view because decisions were made based on consensus and, occasionally, a vote. At first, monthly Exco Meetings took very long because a particular senior individual struggled with collective decision-making, would want to speak on every issue, and would return to it repeatedly – even once a decision had been made.

As was the case with the SEIFSA Council, we had to ask either the Association Manager or the Association Administrator to take minutes of the Exco meetings. It soon became clear to me that it was imperative for me to create the position of a Legal

Services Executive who would double up as a Company Secretary. This made sense since SEIFSA indirectly offered legal services to its members through an external law firm, but merely received a commission from that firm for the business that the federation generated for it.

At the time, more than 65% of the companies affiliated to SEIFSA member associations employed fewer than 50 people. That meant that they were small employers that did not have a lawyer among their staff members. It made sense, therefore, for us to acquire such expertise for the benefit of both the affiliated companies and ourselves: for basic legal advice, we would no longer have to use the services of established law firms but would instead be able to rely on in-house expertise. In addition, as Company Secretary, the Legal Services Executive would also work closely with me to embed good corporate governance at SEIFSA. At our Exco meetings, the new incumbent participated as yet another Divisional Executive, while also assuming responsibility for the recording and production of the minutes, but at the external "Executive Committee" and the SEIFSA Council, she participated only as Company Secretary.

A year earlier, SEIFSA had introduced legal services through the IR Division, which served as a link between the company requiring such services and the law firm used by the federation. Often, the nature of legal services required by affiliated companies related to labour law, hence it was important to ensure that there was an overlap between the IR Division and the new Legal Services Division. The former dealt with IR matters, with the latter handling legal matters, such as representing companies at the Commission for Conciliation, Mediation and Arbitration (CCMA) and the Labour Court and assisting both SEIFSA and affiliated companies with legal compliance.

Once appointed, the Legal Services Executive and Company Secretary also reported to me as Executive Director.

I also ensured that SEIFSA complied fully with the country's anti-smoking legislation. Once properly cleaned and fumigated, the former smoking room was converted into a room where staff members could have their lunch, and smokers were required to indulge in their unwholesome habit on top of the seventh floor of the building, together with those who worked for MIBFA.

A series of annual roadshows to the regions, scheduled for November 2013, had been arranged long before I joined SEIFSA. The federation used these roadshows as opportunities to market itself and to strengthen relations with affiliated companies, which were invited to the functions. However, since the next round of negotiations was a few months away, the focus of the November 2013 roadshows was the 2014 wage negotiations. Companies were keen to hear how SEIFSA expected the 2014 negotiations to pan out. The different Executives and Health and Safety Manager, Nonhlalo Mphofu, took turns to make presentations relating to their respective portfolios on the main anticipated developments in the year ahead, and Trentini – who spoke on the negotiations – was by far the main attraction of the show. For me, the roadshows to Witbank, Cape Town, East London, and Durban were an occasion to be introduced to the federation's key stakeholders.

During those roadshows, it became clear to me that it was Trentini who was SEIFSA's IR guru, and not Angus as IR Executive. Indeed, that became patently obvious in the months and years to come. Whenever there was an IR issue that needed to be dealt with, it was Trentini – and not Angus – who attended to it or assigned staff members to attend to it. Although he held the designation of Operations Director, Trentini did not perform at a strategic level, but worked more as an IR super-executive. Repeatedly, he encroached in Angus's terrain and did not allow him the space to grow into the portfolio or to demonstrate his expertise in it. Over the next few years, I was to repeatedly encourage Trentini to leave IR matters to Angus and to work more closely with me on strategic matters, as his designation required of him.

It was while we were on one of these roadshows that I was to learn something very interesting. From East London, we were due to proceed to Durban for our final roadshow. However, a day before that roadshow was to take place, the KwaZulu-Natal Engineering Industries Association (KZNEIA), to which Duys was affiliated, was due to hold its AGM, and he was keen for me to attend it with Trentini. Consequently, the two of us flew from East London to Durban a day earlier to attend the KZNEIA AGM.

It was while we were driving from King Shaka International Airport to Kloof, where the AGM would take place, that Trentini told me that, sometime in the past, there was a black man of Indian origin – who came from KwaZulu-Natal – who had previously been appointed Executive Director of SEIFSA. I did not know whether that was before or after Carson, and I did not ask. According to Trentini, that man was never equal to the challenge and spent all his time keeping to himself in his office. Three months later, the SEIFSA "Executive Committee" concluded that he did not measure up and parted ways with him. There was absolutely no trace of him at the federation.

"I hope the same will not happen to you," Trentini said, smiling.

I was struck by the information that had just been shared with me. It occurred to me that the organisation must have been much more hostile to change than I had previously imagined, and I wondered whether Trentini had told me the story because a part of him wished, perhaps, that I would have the same ignominious end.

Among the things I did not understand about the structure that I had inherited was the fact that the Executive Director did not have a Personal Assistant (PA). For assistance, I had to rely on a favour by Skills Development Executive Nazrene Mannie, who ensured that an intern in her team, Gugu Shongwe, was often available to assist me when I needed help. I could not conceive of a productive senior executive who would have to make his own drinks, print out or photocopy his own documents, and set up his own appointments. It just did not make sense to me. I had had a PA since 1995 when I was appointed Political Editor on *The Star,* and there I was at the helm of an organisation that expected me to succeed without one.

In mid-December, when he called to wish me a good festive season as he went on holiday, I informed SEIFSA Interim President Henk Duys that I could not work without a PA, hence

I would set a process in motion to appoint one. Subsequently, I assigned Mannie to recruit for a PA for me.

Upon her return from the end-of-year break in January 2014, Venter suggested that I agree to having a new office built for me at the end of the left passage, in the east wing, in what was a communal space that was used as a library. It had ample space, a big bookshelf, a couch and a second television set (the only other one was at reception). After inspecting the space, I agreed that it would make a good office for the Executive Director. Marketing and Communications Executive Adelia Pimentel, whose duties included liaising with the landlord, duly oversaw the construction of the office.

To put an end to the east wing–west wing divide, I also asked Venter to move into an office close to Trentini's. Their offices were adjoined by a third room where a PA who would assist both of them would sit, while the Finance Executive moved into Venter's former office in the west wing.

It was with Mrs Govender (I refer to her in this formal manner because I have since forgotten her first name), who was sourced for me by Mannie, as my Executive PA that I moved to the new office around March 2014. Around the same time, Venter and Trentini employed Monica Pillay as their shared PA. Regrettably, Mrs Govender spent only a few months with me, before she resigned to deal with a family emergency. I replaced her with Lerato Lebeko, who was my PA when I was Public Affairs and Communications Director at Coca-Cola South Africa.

Thus settled, I was ready to turn my full attention to the job at hand.

## Conclusion

Although, in its archived history, SEIFSA painted a glowing picture of its anti-apartheid credentials and its influence on championing change in the M&E sub-sector, the situation that confronted me on my arrival at the federation revealed a picture that could not have been more different. Not only was SEIFSA itself untransformed, but it was also not organised as a modern corporate entity that had to earn its keep. Instead, it was a nepotistic organisation that was not structured for growth.

My first two months in the job were filled with a series of concerning surprises, which I had to tackle head on. Some of the people employed at SEIFSA had worked nowhere else throughout their lives and knew no other organisational culture except the one that prevailed in the organisation, while some had spent only a smidgeon of their early professional lives elsewhere. Although Brian Angus had laid a firm foundation for alternative sources of revenue, SEIFSA was still so heavily dependent on membership fees that it had been reduced to a loss-making entity when the CBL was not renewed following its expiry on 31 December 2012.

SEIFSA, *circa* November 2013, was crying out for change.

# Chapter 5

# Introducing Good Corporate Governance and Implementing a Turnaround Strategy

## Introduction

The strategic review sessions that had begun before I joined SEIFSA continued in November and early December before the office closed for the festive season. I had formed several teams and assigned responsibilities to them ahead of the sessions. Their responsibilities were to meet among themselves, deliberate on the topics given to them, and collectively to take a view that would inform their presentations at the strategic review sessions. Led by different Executives, including Deputy CEO Elsa Venter and Operations Director Lucio Trentini, the teams comprised staff members of various levels of seniority who had been in SEIFSA's employ for years, among them Managers and lower-level employees.

The teams were assigned the following responsibilities: conducting a SWOT (strengths, weaknesses, opportunities, and threats) analysis of SEIFSA; conducting environmental scanning through the use of PESTEL (political, economic, social, technological, environmental, and legal) analysis; and conducting a five-forces (threat of new entrants, threat of substitutes, the bargaining power of suppliers, the bargaining power of customers, and a review of the state of rivalries among competitors) analysis. Another team was tasked with reviewing SEIFSA's positioning statement and developing the federation's Vision and Mission statements, while another still had to propose corporate values to guide the organisation.

One team had to answer the question of what SEIFSA had to do differently and what additional products and services it had to offer, while another was asked to review the SEIFSA structure and to recommend an optimal structure. Venter and I constituted a team whose responsibility was to review the federation's Memorandum of Incorporation (MoI).

The strategy session took place throughout the day at Country Club Johannesburg in Auckland Park on Monday, 25 November 2013.

The respective teams had three weeks within which to prepare for the strategy session. The mandate given to them was clear enough: they had to be independent and courageous in their review of the different aspects of the organisation and be bold enough to make far-reaching proposals that they deemed to be in the long-term interest of SEIFSA and its members. They were encouraged to think out of the box and not to be bound by that organisation's past and culture. Their primary loyalty was to be to SEIFSA, which we all wanted to see surviving into the future, and not to any interested parties.

## Getting the Structure Right

One of SEIFSA's main problems was its loose structure. Seventy years after its formation, the organisation continued to be driven by its member associations, through their formal structure called the SEIFSA Council, with an Executive Director who took instructions from the SEIFSA Council. To make it easy for the SEIFSA Council – which had grown too big and unwieldy – to co-ordinate the federation's affairs, some of its prominent members were elected to what was called the "Executive Committee" so that they could work closely with the Executive Director and oversee his work.

However, the role of this "Executive Committee" was not spelt out in the organisation's MoI. Although legend within SEIFSA has it that the former captains of industry who made it onto the "Executive Committee", like Anglo American Corporation Deputy Chairman Graham Boustred, called the shots and imposed wage settlements on the SEIFSA Council, SEIFSA's MoI at the time was clear that the "Executive Committee" was as an extension of the SEIFSA Council. Its decisions were meant to be recommendations to the SEIFSA Council, which was the body with the authority to make all binding decisions.

One of the consequences of that arrangement was that the leaders of the member associations considered themselves – and not the "Executive Committee" – to be the bosses of the Executive Director. In addition to being entitled to the benefits of SEIFSA's membership, they considered the organisation's Executive Director to be accountable to them. That was an untenable situation not only given the size of the SEIFSA Council, but also because the member associations often took divergent positions on different issues. For example, some associations represent the interests of large employers like ArcelorMittal South Africa, while others represent metal fabricators whose interests are at odds with those of the steel giant. Some associations, like the Non-Ferrous Metals Association, want the export of scrap metal banned, while others either have no position on the issue or are among the loudest proponents of its export.

Therefore, one of my priorities was to get the SEIFSA structure right. While it made sense, years ago, at a time when he had a handful of people reporting to him, to have the person at the helm of the organisation referred to as Executive Director, it did not make sense to appoint two additional Executive Directors (and call one of them "Deputy CEO") and yet still retain that designation. As the benchmarking exercise mentioned in the previous chapter had found, organisations like the Chamber of Mines, Business Against Crime, the National Business Initiative, the South African Chamber of Commerce and Industries, the Consumer Goods Council of South Africa and Business Unity South Africa were led by CEOs, and not Executive Directors. Almost without exception, all of them had only that one person, the CEO, as Executive Director, and yet SEIFSA had *three* Executive Directors who went by the designations of Executive Director, Deputy CEO, and Operations Director, respectively. That did not make sense at all.

In recognition of the fact that there were three Executive Directors and that the number two in the organisation was called the Deputy CEO, it made sense that the most senior position be called CEO (after all, you can't have a Deputy CEO without a CEO!). Next, it was important that a clear distinction be drawn between the structure called the "Executive Committee" and the SEIFSA Council. As a Certified Director (via the Institute of Directors in South Africa) at the time, I was more than familiar with the role of a Board of Directors, something that did not exist in SEIFSA's nomenclature.

Therefore, through suggested amendments to the MoI, I proposed the renaming of what was called an "Executive Committee" to the Board of Directors, with the person elected President of SEIFSA automatically becoming Board Chairperson. Next, we had to accord to the Board of Directors the legal powers accorded to that structure by the Companies Act 71 of 2008, while reserving for the SEIFSA Council the role of being the mandating body on matters of common interest, such as wage settlements and industry matters of mutual interest.

Only the Executive Directors would attend Board meetings, while SEIFSA Executives would continue to attend SEIFSA Council meetings to provide updates on their respective portfolios, respond to any questions that member associations had, or note discussions on matters that fell within their areas of responsibility. The only Executive Committee (Exco) which legitimately continued to exist was the one I had constituted, which was a structure that assisted the CEO in running the federation.

Next, I recommended a reduction in the frequency of SEIFSA Council meetings from monthly to quarterly, like those of the Board of Directors. Provision was made for the Council to meet more regularly when the situation demanded, such as during wage negotiations when member associations would formulate mandates and receive feedback on how things were progressing in the negotiations.

These amendments were all accepted by the Board and subsequently ratified by the associations at the AGM in October 2014.

With the assistance of SEIFSA President Ufikile Khumalo, I developed a Delegation of Authority Framework that was approved by the Board. For the first time in SEIFSA's history, considerable authority was clearly delegated to the CEO for operational matters, and issues that required Board approval were clearly spelt out. Likewise, I delegated authority to the members of my Exco not only to ensure that there was clarity about the levels of authority that Exco members had in their Divisions, but also to ensure that I was not again confronted with the situation which had seen us, in a meeting of the former "Directorate", having to approve staff loan to the value of R12 000. From the May 2014 Board meeting, I started tabling the CEO's Report to ensure that the Board was kept fully informed of developments, including decisions taken by the CEO in the interim on matters delegated to him by the Board.

The Chairpersons of the member Associations also had a structure called the Chairmen's Forum, which saw them meeting on a monthly basis with the SEIFSA leadership at the federation's office. It was a superfluous structure because the same individuals who constituted it were also members of the SEIFSA Council which also met monthly. Since there was no reference anywhere in the MoI (pre- and post-amendment) to the Chairmen's Forum, I had its meetings gradually phased out. Appropriately, the SEIFSA Council became a proper forum for discussion on matters of common interest among member associations and their office bearers.

## A New Reach, Vision, Mission, and Values

It was clear, during the day-long strategic review session, that SEIFSA was a prisoner of its past. Instead of presenting itself as an efficient organisation that was the home of in-demand training and consulting services, SEIFSA had continued to function as an organisation dependent on a loyal and cash-flush membership. The reality, however, was that the poor state of the economy had left the manufacturing sector, and its M&E component, severely limping. Not only was the number of companies affiliated to member associations progressively declining, but there was also a downward trend in the take-up of the federation's products and services by these companies.

The organisation was not aggressive in its marketing efforts, and it did not compete in the market for ideas or the best available talent. "Providing industry solutions" was its uninspiring slogan – I have no idea for how long the federation had had it.

One of the things that became clear during the strategy session was that SEIFSA needed to consider the entire M&E sub-sector as its market and audience, and not only those companies within the industry that were affiliated to associations which were members of SEIFSA. Indeed, the declining uptake of SEIFSA's products and services made it imperative, as Nazrene Mannie suggested, that the federation marketed and offered its products and services to anyone who wanted them – including entities beyond M&E. There was a lot of merit in that suggestion, which was duly accepted.

After all, it is not only companies affiliated to SEIFSA member associations, or those that are in the M&E sub-sector which dealt with IR challenges. Services such as chairing disciplinary hearings or representing companies in such disputes at the CCMA are required by a far bigger universe of companies. It thus made sense to market such services beyond the M&E sub-sector. Similarly, it was all industrial companies, and not only those in the industry, that could take advantage of the apprentice and artisan training services offered by the federation through the Fundi Training Centre (which I renamed the SEIFSA Training Centre). The same went for most of the other products and services. In particular, the *SEIFSA Price and Index Pages* (*PIPS*) were long in demand across various economic sectors.

With a growing number of companies in the industry establishing a presence – either as traders or as manufacturers – in the Southern African Development Community (SADC), there was an opportunity to extend SEIFSA's reach to the SADC region. Accordingly, we recommended the establishment of SEIFSA chapters or member associations in the region, and the change of SEIFSA's name to Steel and Engineering Industries Federation of Southern Africa, and not just "South Africa". This recommendation was also accepted, and we went on to register

SEIFSA in Zambia and Namibia and embarked on a process to get the Engineering Iron and Steel Association of Zimbabwe to take membership of SEIFSA (that attempt fell through when some of the people in the leadership of the association in that country developed cold feet).

Following the presentation made by the team which had been tasked with considering the issues, we eventually settled on the following Vision: "To promote sustainable metals and engineering industries to ensure that they are strategically positioned for innovation and growth in the interests of a prospering South Africa". We adopted a two-fold Mission statement, which was captured as follows: "To be Southern Africa's most respected advocate for the metals and engineering industries in order to create innovative businesses positioned for growth and working in partnership with all stakeholders in the interests of South Africa. To foster mutually beneficial relationships between employers and labour in the industries and to help members develop their human capital to realise their full potential".

Collectively we adopted six Values, each accomplished by a short explanatory statement. These are listed below.

### Integrity

Integrity is paramount to us. It informs everything that we do and how we do it.

### Diversity

We embrace, value and leverage Diversity.

### Excellence

We seek to do everything right the first time, with Excellence.

### Stewardship

We take Responsibility for our actions and treat SEIFSA's assets with respect.

**Passion**

We approach every task, however small it may appear to be, with Passion.

**Innovation**

We always strive to Improve our performance and to come up with new products and services.

Although this was a team effort, I took an active lead in the development of the Vision, Mission and Values and assumed responsibility for crafting them.

## The 2014–2017 Strategy

During the strategic review session, it was obvious that the biggest challenge for which we needed to find a solution was SEIFSA's deepening financial crisis. We needed to arrest the financial decline and return the organisation to a sustainable path. That meant that we had to find ways to grow our revenue, while keeping a watchful eye on costs. At the time, our emphasis was disproportionately more on revenue generation, rather than on cost containment.

That forced us to embark on a careful review of the federation's pricing strategy. Were SEIFSA's products and services appropriately priced? Could we, by reducing the prices of some products or services, sell more of them and, in the process, generate more revenue? Were there products and services whose prices we could adjust upwards without further losing market share?

There was not much wiggle room. Already, there was a discernible downward trend in the uptake of SEIFSA's products and services by member companies. We needed to entice more companies, inside and outside the M&E sub-sector, to start using SEIFSA's products and services, and had to guard against doing anything that would further put them off doing so. When we compared our prices with those charged by competitors for the same services, we concluded that, although there were limited instances when SEIFSA was more expensive, generally the federation's products were competitively priced.

However, at least one product, which was unique to SEIFSA, was so cheaply priced that it was almost a freebie. That was the *SEIFSA Price and Index Pages* (*PIPS*), a then 55-year-old, subscription-only monthly magazine that published "indices for material and services costs, statutory and actual labour costs and a number of sector-specific indices developed for use in the metal and engineering industries," in the words of the federation's EC Division. It was a niche publication to which non-member companies in diverse economic sectors, including mining and construction, subscribed – and it was exclusive to SEIFSA. In *PIPS*, SEIFSA had a unique product, and yet it was not priced as such.

As I became better acquainted with the publication and its uniqueness, I realised that it presented us with a *sui generis* opportunity to generate more revenue. As I recall, at the time, an annual subscription was priced around R4 500 for all companies, regardless of whether they were members of affiliated employer associations. I suggested that, at the beginning of the new financial year on 1 July 2014, we adjust the annual *SEIFSA PIPS* subscription rate upwards by a hefty 60%, with the blow cushioned for member companies by extending a 40% annual discount to them. That way, *SEIFSA PIPS* would be an important membership recruitment tool: companies which wanted to enjoy the 40% discount would have to take membership of a SEIFSA member association or, if they were not in the M&E sub-sector, they would have to take advantage of a direct SEIFSA associate membership provided for in the organisation's MoI.

Since *SEIFSA PIPS* is collated and published by the EC Division, the repricing immediately made the EC Division the largest in terms of revenue generation. This was in addition to revenue it generated through training and consulting on *PIPS*, the economics of the M&E sub-sector and the calculation of Contract Price Adjustments.

Banking on the change from a lackadaisical to a more business-focused organisational culture, all divisions were required to prove their worth to the federation by generating as much revenue as possible to cover their respective costs, at the very least, but ideally to generate profits. Everybody needed to earn their keep. I stressed on every occasion that I had that SEIFSA would no longer carry passengers. The culture I wanted inculcated was that of a professional organisation akin to a consultancy, one which valued every minute of its time. We began to monitor divisional budgets – and performance against those budgets – more rigorously.

To show just how much at home employees felt at SEIFSA, nobody had any performance contracts in place, and the organisation had no culture of performance appraisal. Instead, its remuneration structure included not only a guaranteed thirteenth cheque in December, but also a guaranteed performance bonus at the end of the financial year in June. Thus, although SEIFSA had made a loss the previous year and had budgeted for a loss in the 2013/14 financial year, staff members fully expected to receive their bonuses when the year ended. The whole thing was devoid of business sense.

Therefore, with the reluctant support of the Executive Committee, I announced that there would be no bonuses paid at the end of that financial year and that, in the new financial year, everybody would be required to conclude a performance contract with those to whom they reported. I led this process by developing corporate goals for the 2014/15 financial year and sharing them with the Executive Committee, which had the opportunity to provide input. When that was done, I asked the Executives to develop their own performance contracts, which were presented to all of us for calibration. My own performance contract, which was subsequently approved by the Chairman of the Board and the Remunerations Committee, was a summation of the Executives' performance contracts.

Each performance contract had five aspects to it: Living the SEIFSA values, financial targets, strategic targets, customer goals, and human capital deliverables. Assessment of living the values was based on a 360-degree employee survey, while assessment of performance in relation to customer goals was based on a survey of internal and external customers.

For us to be able to attract more companies within and beyond the M&E sub-sector in terms of the use our products and services, we needed to *demonstrate* externally that we had the requisite expertise within the federation. Therefore, as part of their performance contracts, all Executives were required to publish a minimum of four opinion pieces per year on their respective areas of expertise in the mainstream media. Throughout my stay at SEIFSA, this was a requirement that most of the Executives struggled with, except for the various Chief Economists that we had. In particular, Dr Michael Ade excelled in this regard and often published more than the required minimum number of opinion pieces.

Except for Ade (and Henk Langenhoven before him), most of the executives wrote very poorly. I was amazed by how university graduates, including those whose first language was English, struggled to write sensibly. They also struggled to write comprehensive submissions for our Executive Committee meetings. Even with the help of the Communications Manager, among whose duties it was to help them to craft or polish their opinion pieces before submitting them to me for editing once she was happy with them, some of the Executives considered themselves to have done well if they managed to get one opinion pieces published in *The Star*, *Sowetan*, or *Business Day* in a year.

To accomplish our goal of attracting more companies to using SEIFSA's products and services, we committed to a growth strategy. We undertook to establish a Small Business Hub within the EC Division through which services specifically targeted at relatively small companies would be curated and offered. That needed the services of a Commercial Manager, who would report to the Chief Economist, so one was employed. Venter had come up with the idea of a Small Business Hub, as part of her responsibility for business development.

Among the main elements of the three-year turnaround strategy were the following:

· Broadening SEIFSA's geographic reach to include countries in the SADC region;

- Setting up the technological capability to respond to calls for advice or help from members in any one of the areas in which SEIFSA offers a service or product;
- Identifying, developing and marketing new income streams in the form of products and services and pricing them appropriately;
- Linking SEIFSA with international fraternal organisations with the ability to supplement the federation's expertise;
- Establishing the capacity for SEIFSA to be able to make direct representations on proposed legislation with the potential to impact on the M&E sub-industries;
- Developing economic research and development capability, internally within the EC Division or through a partnership with a credible third party;
- Leveraging information and communication technology to offer e-training and, in the process, generate revenue;
- Actively marketing the SEIFSA Training Centre and leveraging its reputation as one of the top artisan training institutions in the country;
- Replicating, through acquisition or a joint-venture partnership, the SEIFSA Training Centre in two to three of the major industrial centres in the country;
- Establishing SEIFSA as the first port of call for information on manufacturing, particularly on M&E industries, in southern Africa; and
- Entering into strategic, win-win partnerships, including possible joint ventures, with institutions or organisations capable of adding value to SEIFSA's products or services.

Both the three-year strategy and the new structure were approved by the Board of Directors at their meeting on 3 February 2014. However, the Board peremptorily rejected what I had considered a vital part of the strategy: the establishment of a for-profit SEIFSA subsidiary which would operate like any company and pay tax, with its profits ploughed into SEIFSA as dividends, to subsidise companies' fast-rising membership fees. The proposal's rejection robbed SEIFSA of a potentially reliable source of revenue to diversify its income streams and meant that financially SEIFSA would continue to be dependent on a mixture of membership fees and on revenue generated through its products and services. With revenue from membership fees on a downward spiral, with the possibility of hiking such fees being slim, that meant that going forward there would be great pressure on SEIFSA to earn its own keep.

To deliver on this strategy, we needed to supplement the institutional knowledge which resided at SEIFSA with new energy and expertise. In particular, I needed people on that Executive Committee who would serve as allies – in addition to Marketing and Communications Executive Adelia Pimentel and Skills Development Executive Nazrene Mannie, who had eagerly bought into the vision of a new SEIFSA and already played that role. Early on in my tenure, I changed Mannie's designation to Human Capital and Skills Development Executive and placed human resources (which until then had been handled by MIBFA on SEIFSA's behalf) under her. When she was subsequently recruited by Angela Dick to join Transman, she was replaced by Mustak Ally.

I appointed Bridgette Mokoetle (nee Mphuthi) as Legal Services Executive and Company Secretary and Rajendra Rajcoomar as CFO. The latter position became vacant when former Finance Executive Nicolene Harmse resigned for health reasons. Both Mokoetle and Rajcoomar came with solid academic qualifications and enviable experience in their respective areas. I also promoted Nonhlalo Mphofu from Health and Safety Manager to Safety, Health, Safety and Quality (SHEQ) Executive, which saw her joining the Executive Committee.

When I told him of my plan to promote Mphofu to an Executive position, Trentini strongly motivated for me to move Tafadzwa Chibanguza, a Zimbabwean young man who was studying part-time for a Bachelor of Commerce Honours degree at the University of the Witwatersrand, from being a Junior Economist to the rank of a Manager. Trentini spoke very highly of the young man who had joined SEIFSA in January 2013, whom he considered a potential successor to Langenhoven in years to come. After checking with Langenhoven, I duly promoted Chibanguza to the position of Economist, which was at the same rank as that of a Manager.

In a matter of months, this saw the Executive Committee evenly split racially and more representative of the country in which SEIFSA operated: Deputy CEO Elsa Venter, Operations Director Lucio Trentini, Chief Economist Hendrik (Henk) Langenhoven, IR Executive Gordon Angus and Marketing and Communications Executive Adelia Pimentel were the white colleagues, while Mannie, Rajcoomar, Mphofu, Mokoetle and I were the black members of the team. Pimentel and Mannie readily embraced the vision for a new SEIFSA, Langenhoven was a true professional who stayed above corporate politics, and Trentini adopted a wait-and-see attitude, while Venter appeared to be opposed to the direction in which I was leading SEIFSA.

For the Legal Services Executive and company Secretary position, I asked Trentini to join me and Mannie as we conducted interviews, and, for the CFO position, I asked Venter to join me. My initial intention was to appoint a white man to the CFO position so that I could manage unfounded fears – which had begun to gather momentum – that I was set on turning SEIFSA into a black organisation. Thankfully, Aveng Grinaker Finance Executive Craig Vernon Werner emerged from the interviews as the best candidate. After he had completed our standard psychometric assessment, I offered him the job through our recruitment company – but he subsequently turned us down.

That was when I then offered the job to the second-placed candidate, Rajcoomar, who was then CFO at the FoodBev SETA. He was a solid candidate with a Master of Business Administration degree from the Wits Business School who had described himself during the interview as being "loyal and dedicated", as having "a thirst for knowledge", and as being "a good team player". Over the years that I worked with him, he proved to be all those things – until in my last year with the federation.

Pimentel, Mannie (and later Ally), Rajcoomar, Mokoetle, and Trentini were reliable colleagues who either readily or subsequently fully bought into the vision for a new SEIFSA. It is thanks to their work and dedication that we made good progress in implementing our turnaround strategy.

To ramp up our communication efforts, I created the position of Communications Manager within the Marketing and Communications Division. Mannie sourced candidates from a recruitment agency. Once they were done with the shortlisting, she and Pimentel conducted the interviews and Pimentel recommended Aurelia (Ollie) Madlala, whose appointment I duly ratified. From then onwards, we raised SEIFSA's public profile considerably through the number of press releases that we issued and the radio and television interviews that we regularly conducted.

For instance, in May 2014, when President Jacob Zuma announced his second Cabinet, I was the first person, through a SEIFSA press release, to condemn the size of his Cabinet. I pointed out that Nelson Mandela's and Thabo Mbeki's Cabinets had 28 Ministers and that Zuma had pushed that number to 34 in 2009 and 35 in 2014. That, I argued, compared very unfavourably with First World countries like the USA, Japan and Germany, which had 15, 18, and 16 Cabinet Ministers respectively.

"That is very, very large. Much of the money that will be going to paying the Ministers' and Deputy Ministers' salaries and paying for their lifestyles should be redirected to service delivery," I added during an interview with Radio 702 (Eyewitness News, 2014).

## Conclusion

The 2014–17 strategy – which was developed jointly with SEIFSA fellow Executive Directors, Executives and Managers – laid a very firm foundation for the organisation's eventual turnaround from a loss-making entity to a profitable one. To this end, the solid support of the Board of Directors of the time was crucial.

Given the structure of the federation and the fact that some within it had been in its employ for decades, they enjoyed long-standing, mutually beneficial relationships with people in leadership of the affiliated associations. That meant that they had easy access to influential individuals in these associations, to some of whom they embarked on a whispering campaign against me as CEO. In a relatively short space of time, much had changed within SEIFSA – and even more change was looming. Knowing how uncomfortable change was for some people and how difficult they found it to embrace it, neither I nor the Board was surprised by this development. The important thing was that there was a job to do, and the Board and I were committed to that project.

I was to note in a memorandum to the Board in April 2014, on the review of my six-month probation in the job: "Change, it is a well-known fact, is not easy, especially when people have been at a place for many years, have done things in the same way throughout those years, and have cultivated strong ties inside and outside the federation. Many of them have considerable influence with some people on the Board and in some associations."

Having listed my objective achievements in the job in my first five months, I concluded my memorandum thus: "It should be clear, from the achievements mentioned above, that nobody who is unprejudiced can deny that, in my first five months in the position, I managed to lay a very solid foundation for a better-known and more influential Steel and Engineering Industries Federation of Southern Africa, which is poised for growth. Therefore, there can be no gainsaying that, the whispering campaign notwithstanding, if the first six months of probation were a test, then I have passed that test with flying colours."

The Board agreed – and confirmed me in the position.

However, the fight was far from over. Like the broader manufacturing sector, the M&E sub-sector was bleeding at the time. To help you, the reader, to appreciate this better, the next chapter focuses on the views of several stakeholders to whom I spoke at the time.

# Chapter 6

# An Industry in Deep Trouble

## Introduction

Although the immediate impact of the GFC of 2008/9 on South Africa's metals and engineering (M&E) sub-sector was relatively benign, the aftermath of the crisis was devastating. More than a decade later, the sub-sector has not yet recovered to pre-GFC levels. During this time, a number of companies were liquidated and many others collectively laid off thousands of employees.

The M&E sub-sector, especially the construction industry within it, benefitted handsomely from the massive build programme in the run-up to the 2010 FIFA World Cup in South Africa. The amount of public infrastructure investment made ahead of that global event initially cushioned the M&E sub-sector, but after completion of the stadia and related infrastructure, it was left similarly exposed to the vicissitudes of the economic climate.

The 2013-2016 period was particularly harsh for the sub-sector, which has a symbiotic relationship with the mining, construction, and auto-manufacturing industries. Not only the construction sector, but the mining sector too struggled during that period. For instance, year-on-year fixed investment in the mining sector dipped to 5,5% in 2010 and grew by only 1.7% in 2013, down from growth of 4.3% in 2012 (Chamber of Mines, 2013). There was great volatility in the mining sector: value add increased by 4% in 2013, decreased by 1.4% in 2014, grew by about 4% in 2015, and again decreased by 4.7% in 2016. The same trend was evident in mining companies' share of profits— annual reported profits before taxes declined by 60% in 2013, improved by 109% in 2014, grew by about 4% in 2015, and improved by a whopping 255% in 2016. While gross fixed investment declined by 1% on average from 2013 to 2016, the net investment worsened progressively: -2.5% in 2013, -0,4% in 2014, -10,5% in 2015 and more than 100% down in 2016 (Chamber of Mines 2016).

In construction, the shortage of major infrastructure projects led to established companies like Basil Read, Group Five and Esor Construction being placed in business rescue in 2018 and 2019 (Zingoni, 2020).

As was indicated in chapters 1 and 2, both the manufacturing sector and its M&E sub-sector had received tremendous support from the National Party government in years gone by. Not only had they received subsidies and enjoyed protection from international competition through the imposition of import tariffs, they also benefited from massive infrastructure projects led by some of the country's SOEs. However, that changed when democracy dawned and import tariffs were prematurely removed overnight, without local manufacturers having been given a period of respite to prepare themselves for entry into the global market.

By the end of 2013, the country's SOEs were in the grip of President Jacob Zuma, and a growing number of them were not in a position to engage in infrastructure projects of the kind which had previously powered local manufacturing. That was the state of affairs when one joined SEIFSA in November 2013. There was a lot of disaffection among many employers, and they held the Zuma Government directly responsible for their plight. Although also concerned about a non-sympathetic Government which appeared to have other priorities, labour was far from understanding. This chapter gives expression to some of those frustrations.

## "We are Not Able to Compete"

In the early months of my tenure, among my priorities as CEO was to get to know the leaders of our member associations. I realised that, since there were 27 member associations, it would take me long to have individual meetings with each of them. With this in mind, I asked for a list of the leaders of the bigger associations and made appointments to meet with them.

One of my first meetings was with Rob Pietersma, CEO of CBC Fasteners and chairman of the South African Fastener

Manufacturers Association (SAFMA), on 2 December 2013 in his office in Krugersdorp. Fasteners, Pietersma explained, are used in the mining process. He told me that National Bolts, which was once the largest manufacturer of fasteners and bolts, had merged with CBC Fasteners in 1998 and that the number of companies in the sub-industry had declined from 22 to 16. These were the companies that constituted SAFMA. CBC Fasteners was the largest among them.

Pietersma had started representing CBC Fasteners in SAFMA since he became CEO of the company in 1996 and had been attending SEIFSA Council meetings since 2007. SAFMA's primary interest, he explained, was in SEIFSA's wage negotiations and in the federation securing Government support for the fastener industry in the form of import tariffs.

When I met Mcebisi Jonas, then the Eastern Cape Member of the Executive Council for Economic Development in Bhisho two days later, he blamed the Government's Motor Industry Development Programme (MIDP) for the poor state of manufacturing in the province. He argued that the MIDP was part of the problem because original equipment manufacturers (OEM) based in the province had become part of their companies' international supply chains. Consequently, he said, "all manufacturing, outside of auto [manufacturing], is gone". He felt that the Government had to find ways to incentivise local manufacturing.

At a meeting of SEIFSA's Metals Fabrication Cluster on 22 January 2014, a participant pointed out that, in the 1960s-70s, when there was a 100% tariff on imported vehicles, the then government required that 70% of the components of vehicles manufactured in South Africa had to have 70% of their components manufactured locally.

When I met him on 10 December 2013, Johann Ellis, Chairman of the Electrical Manufacturers Association of South Africa (EMASA), told me the association was born 12 years earlier as a break-away from another SEIFSA-affiliated association, the Electrical Engineering and Allied Industries Association (EEAIA), which was more focused on importation. The majority of EMASA's members continued to focus on manufacturing, although some of them had since begun to import.

Asked what he would do differently if he ran SEIFSA, Ellis replied: "I would improve the status of SEIFSA within the industry and the business community. We don't have the voice the Chamber of Mines has. I think we need that."

The meeting with Bob Stone, ZIMCO Executive and Chairman of the Non-Ferrous Metal Industries Association (NFMIA), was the first to alert me to the challenges experienced by some of the companies and sub-industries in the M&E sub-sector. Although he was a strong supporter of the federation, Stone nevertheless held the view that SEIFSA was "not supportive at all" when it came to his association's concerns about the continued export of scrap metal," and said the organisation was "known to be focused on HR/IR".

"The scrap metal merchants charge some of the highest prices in the world for scrap metal," Stone said.

At an MEIBC Job Creation Summit in Boksburg on 27 January 2014, Karl Cloete, then Deputy General Secretary of NUMSA, noted that, in Ghana, the export of scrap metals was a crime. Although that was the Ghanaian government's intention at the time, in mid-April that year, Ghanaian Trade Minister Haruna Iddrisu announced that the export of ferrous scrap metal was banned with immediate effect (Reuters, 2014).

Stone told me that, in its heydays, ZIMCO, an alloys manufacturer, produced 900 000 tons of alloys for the local market and 300 000 tons for the export market. However, in 2013, the company had produced a paltry 50 000 tons of alloys. He said that had it not been for its contract with BHP Billiton, ZIMCO would not have survived. He added that the company used to have 65-70% capacity utilisation of its factory, but that had declined to 30%.

In our meeting in Sasolburg on 13 January 2014, Peter Viljoen, Chairman of the Pressure Equipment Manufacturers Association (PEMA), complained bitterly about the effect of imports on members of his association. He said there was a need for big companies, which were not covered by the Government's designation of pressure equipment, to be persuaded to buy it from local manufacturers.

Viljoen bemoaned the fact that artisans did not enjoy much respect in South Africa. He said it was important for the country to acknowledge artisans' importance as professionals in the way that Germans – who develop artisans to be Master Craftsmen – have done and to pay them accordingly. He also complained about the fact that, although South Africa boasted a large engineering industry, "all the engineering is done by foreigners".

In my meeting with him on 14 January 2014, ABB CEO and EEAIA Chairman Leon Viljoen said the association comprised local and international companies, including importers and domestic producers of electric motors and refrigerators, among other things. He shared the concern raised by Ellis, Stone, and others that SEIFSA was seen as "just an IR federation", and said it was that perception which had led to "mostly HR persons participating in the federation".

Viljoen told me about a Chinese company which had approached ABB to partner with it in tendering for work at Transnet. He said when the company won the tender, it dropped ABB. Localisation, he said, was much talked about, "but we don't see it happening".

In a meeting later that that day, Aberdare Cables Executive and Association of Electric Cables Manufacturers of South Africa (AECMSA) Chairman Keith Edmond also stressed the importance of designation as an instrument to get SOEs and the public sector to use locally produced products and spoke about the importance of ensuring that the association's members have ready access to copper. He revealed that Aberdare Cables used to produce 140 000 tons of copper rod but had recently been reduced to producing a mere 40 000 tons. The company, he said, found itself in a position where it had to import 55% of its copper.

"South Africa imports all of its aluminium, and the country will soon find itself importing all of its copper," he warned.

Edmond felt strongly that the Government should incentivise the manufacturing of aluminium rod in South Africa. He explained that, since copper and other metals were listed on the London Metals Exchange, South African companies bought them at the international rate and had to pay further for their conversion into rod.

When we met in my office on 22 January 2014, South African Valve Actuators Manufacturing Association Chairman Greg Walker told me that the local industry was heavily reliant on designation and protection through import tariffs. Designation alone, he said, was not effective, pointing out that although valves were designated, in the previous two years, local manufacturers had lost out on work valued at around R2 billion.

"For every R1 million spent by an SOE with a local manufacturer, a job is created within the sector. We need that designation to work", Walker said.

According to him, in 1994 local valve manufacturers employed about 3 500 people, but at the beginning of 2014 they employed "fewer than 800 people". Walker held the National Treasury partly responsible for this situation. He said the Department of Trade and Industry had approved valve designation on 30 January 2013, but the National Treasury had taken time to sign off on it, "so SOEs ignored it".

In a meeting on 5 February 2014, Rick Allen of Meshco alleged that "not a single member of SAISI ([the South African Iron and Steel Institute] is making a profit in South Africa." The problem, he said, was that although it was a developing country, South Africa applied "First World competition legislation". He said the entire steel industry was "in a total mess", and yet companies could not talk to one another. Allen revealed that in 2008 the Meshco group of companies employed 2 000 people, but by February 2014 that number stood at 1 000. He argued that it was important for clusters of industries in distress to be designated *in toto*.

When I met with him on 14 April 2014, Robert Wilmot, Executive Director of the Hot Dip Galvanisers Association of South Africa (HDGSA), expressed concern that "all zinc in South Africa is now imported". He felt that it was important that the Government came up with plans to monitor the zinc price closely.

While the concerns raised by these leaders touched me, I was deeply moved by the *cri de coeur* by Mike Klein, a company owner and Chairman of the Hand Tools Manufacturers Association (HTMA). After the meeting, I published an impassioned opinion piece in *Business Day*, calling – yet again – on the Government to heed the calls of local manufacturers.

I met Klein at his factory in Randburg on 25 March 2015, after the month-long strike of July 2014 following an impasse in the wage negotiations in the sub-industry. His biggest concern was that his industry was shrinking fast as companies are shutting down because of imports. He accused the Government of not appreciating the fact that manufacturing was about output volumes.

He explained: "My policy is if we can manufacture it competitively, we manufacture it. If we can't, then we import. We are currently manufacturing 40% of our products, and we import the remaining 60%. It costs me R2.65 to manufacture something, yet it can be landed here from Asia at less than half the price, e.g. 93c. Our biggest problem in South Africa is the volume. The unit costs are high because there is no ability to use the facilities 24 hours a day. We have 33% capacity utilisation at the moment".

Klein complained about the cost of raw materials and labour, and the fact that local manufacturers had to meet "rigorous SABS standards, but imports are not of the same standards".

"What do you do?" he asked rhetorically. "You've put all your money into the business and you've got your back to the wall."

Klein, who had been in manufacturing throughout his life, had inherited a family business. He said the strikes in the platinum-mining industry, his primary customer, and that in the M&E sub-sector the previous year had hurt his business badly. He remained sanguine, however, that not all was lost in South Africa, provided that all stakeholders worked together.

A year later, when I met him on 16 May in Wadeville, Germiston, Genrec Engineering Managing Director Fergus Derwin was very concerned about the state of the economy in the country.

"The industry is bleeding," he said. "There is no work in South Africa. International companies want nothing to do with Genrec because of labour instability. The price of steel is also a problem."

Derwin revealed that, owing to high administered and logistical costs inland, Genrec was considering relocating to Durban to be close to the export depots. He said his company had cut its number of employees from 3 000 to 600 – "and we are still looking to cut further".

## The Government's Response?

Presented above is a sample of some of the concerns that were raised by key employers and industry bodies in the M&E sub-sector. Times were tough and, for many in manufacturing, it felt like they had been deserted by the Government. Did Pretoria care about their feelings? If so, how did it respond?

Early on in my tenure, I wrote to Dr Rob Davies, Minister of Trade and Industry, asking for a meeting on behalf of SEIFSA. To his credit, Davies offered us a meeting in Cape Town on 11 February 2014. I took Chief Economist Henk Langenhoven with me to the meeting and asked three Board members – Cape Town-based Norbert Claussen, Durban-based Henk Duys and Johannesburg-based Alpheus Ngapo – to accompany us. Langenhoven and I had prepared thoroughly for the meeting and had shared our presentation with the Board members.

In this section, I reflect on the Government's response, through Davies, to the concerns raised by the sub-sector and on the responses of members of the African National Congress's Economic Transformation Committee (ETC) with whom we had meetings.

In his response to our presentation, Davies said that he was aware of the distress facing the sub-sector, but that things were "looking up" because buses and train carriages were now being manufactured locally by the likes of Scaw Metals. He said the Government was willing to raise tariffs up to the World Trade Organization's bound rate "on the basis of evidence to support industrial development". He said that he was aware that dumping of inferior products was taking place in the country, but urged SEIFSA to take on the role of policing such imports and of reporting to the International Trade Administration Commission (ITAC), which is part of his Department.

"Let's develop a stronger partnership to deal with this matter," Davies said.

The Minister said the Government was keen to know of all instances in which designation was not followed by the public sector and SOEs. However, he acknowledged that, under the World Trade Organisation's trade investment protocol, Pretoria could not force the private sector to support designation. The best that they could do was to conclude social compacts with the business sector, even though these were often ignored.

Trade and Industry Deputy Director-General for Industrial Policy Garth Strachan, who was the only other person with the Minister in the meeting, claimed that there had been "very good cooperation from Eskom and Transnet on designation". There was no doubt, he said, that "at Government level there is commitment from all of us to local procurement". He added that the Government was also determined to ensure that beneficiation occurred in the local economy to create a "real competitive advantage for manufacturing". He stressed that the Government would go ahead with shale gas exploration and production because that would have the effect of lowering administrative costs.

Minister Davies said the fact that the rest of the African continent was becoming industrialised meant that South Africa had to prepare itself for more competition.

In meetings with it, the ANC's ETC generally shared our concerns about the state of the economy in general and the M&E sub-sector in particular. The committee's chairman, Enoch Godongwana, had the advantage of previously having been General Secretary of NUMSA, so he was very familiar with the industry.

At one of the meetings, Godongwana shared his personal view that ideally South Africa should have a tax incentive for companies which invested in labour-intensive industries. He also shared our concerns about the slow pace of public infrastructure investment. What was needed, he said, was "massive investment in infrastructure" to kick-start the economy. He also advocated the tightening of the legislative framework to promote localisation and spoke favourably about a need for "a focus on beneficiation of minerals".

Notwithstanding the generally positive sentiments of policy makers like Minister Rob Davies and others, not much changed on the ground. Instead, the long-standing concerns of companies and their leaders in the M&E sub-sector continued to be heard – and have been raised to a crescendo.

## Conclusion

Like the rest of the South African economy, the M&E sub-sector of manufacturing continued to struggle since the global financial crisis of 2008/9. Sadly, various concerns raised repeatedly by organised business bodies like SEIFSA fell on deaf ears. It was as if the country's leadership had other priorities, with business at the bottom of that list. Inevitably, in the post-2009 period, a significant number of companies went out of business or merged with others to ensure their survival – and hundreds of thousands of jobs were lost as retrenchments became a common occurrence.

As the economy worsened, so, too, did relations between employers and labour. The latter approached negotiations with a view to extract as much benefit as possible for its members, even if that led to some of those members subsequently getting laid off, and business fought a rear-guard battle to limit the effect of yet another increase in an input cost.

# Chapter 7

# The 2014, 2017, 2020/21 and 2024 Wage Negotiations

## Introduction

At the beginning of 2014, the then little-known Association of Mines and Construction Union (AMCU) led 70 000 workers on a strike in the platinum sector. Led by Joseph Mathunjwa, AMCU demanded that its members' salaries be more than doubled from a minimum of R5 000 a month to a new minimum of R12 000. Employers, on the other hand, dismissed AMCU's demand as unreasonable and offered to adjust salaries by 10%.

That impasse led to the longest – and probably most violent – strike in the country's history, certainly since the dawn of democracy. Having begun on 23 January 2014, the strike dragged on for five months, ending on 23 June. In its wake, Impala Platinum, Amplats and Lonmin collectively lost about R24.1 billion as a result of 440 000 ounces of platinum being taken out of production. It was during that strike that South Africa was to witness its mass killings in the democratic era, the Marikana massacre (South African History Online, n.d.).

Even as M&E employers and unions finalised their respective plans for the 2014 wage negotiations, the AMCU strike was raging on in the platinum mining sector. Just as NUMSA followed AMCU's bravado during that strike, employers kept an eye on the response of the three platinum majors.

## A Hardening of Attitudes in 2014

Collective bargaining in the M&E sub-sector takes place in the Metal and Engineering Industries Bargaining Council (MEIBC), a neutral body accountable to both employers and organised labour in the industry in the form of a structure called the Management

Committee (MC). The person running the institution, (called the General Secretary, is appointed by the MC and is responsible for running the council's affairs. That includes convening meetings of the MC and its sub-structure, the Finance and Administration Committee, and providing secretarial services to them.

Money permitting, the MEIBC has traditionally arranged pre-bargaining conferences for the protagonists involved in the talks to offer them an opportunity to weigh up one another's positions and to signal their respective approaches to the forthcoming negotiations. In 2014, the pre-bargaining conference took place in Benoni on 26 March. At the end of it, the labour-employer protagonists officially send one another their respective demands, through the council, and that is followed by the first official round of negotiations. Historically, the main players on either side were SEIFSA, on behalf of its member associations, and NUMSA. Once these two parties had reached an agreement, all the other unions (and independent employer associations) in the MEIBC generally went along with it.

However, the NEASA's registration with the MEIBC as an employer party had considerably muddied the waters. With a number of smaller employers in the sub-sector belonging to it, NEASA became a not-insignificant voice in the affairs of the MEIBC's MC. Over time, the organisation, which was not known for its attachment to collective bargaining, questioned and attacked SEIFSA's *raison d'être*, arguing that the federation had no right to participate in the MEIBC because it was a federation of employer associations and was not itself a registered employer's representative.

There was no love lost between the two organisations. Their respective leaders tolerated each other, although the mutual antipathy was generally known to run deep. Matters were not made easy by the strong differences in their respective ideologies and approaches to things. Where SEIFSA was a long-standing proponent of collective bargaining, NEASA hated it, preferring shop-floor bargaining. Where SEIFSA had historically cultivated good relations with labour, NEASA disliked labour, especially the radical NUMSA. Where SEIFSA's disposition was to work with the democratic Government and to seek to persuade it to do the right thing, NEASA's disposition was to hurl insults at the Government, whose leaders it despised and whose labour laws it strongly detested. During my tenure as CEO, SEIFSA's philosophy was that the Government and labour were indispensable partners with whom we had to work cooperatively in the best interests of the country, while NEASA had a low opinion of both of these stakeholders and considered them a necessary evil.

On the labour side, the only other player of some significance was Solidarity, which represented white – and predominantly Afrikaner – workers, mainly artisans. On its own, Solidarity could not sway the outcome of the negotiations with employers but, in partnership with NUMSA, it was a player not to be ignored. That would explain why SEIFSA Operations Director Lucio Trentini had, before the start of the negotiations, introduced me to each of the two unions' chief negotiators. Although less dogmatic and political than NUMSA, Solidarity was more constructive – some would even say more progressive – than NEASA in the negotiations.

As it turned out, the pre-bargaining conference in March 2014 was more of a shadow-boxing session. Neo Bodibe, NUMSA's Head of Policy and Research Institute, sketched a situation in which workers were struggling to survive because of the state of the country's economy. In that situation, she argued, it would be important for employers to alleviate their workers' conditions by agreeing to a 20% increase in wages and by providing housing allowances and medical aid. Chief negotiator, Vusi Mabho, later continued in the same vein, saying how employers had a duty to assist their employees to cope in the tough economic climate. In his presentation, Solidarity's Francois Calldo painted a picture of life having become unbearable for workers because of the increase in the cost of living. He went through a list of items – mostly necessities – the prices of which had recently increased and complained about the effects of rising administered costs, such as those of electricity and the toll roads in Gauteng.

When it was our turn to speak, Economist Tafadzwa Chibanguza, who was standing in for Chief Economist Henk

Langenhoven, stressed that the conditions which so adversely affected employees also had the same effect of companies within the SEIFSA stable. "We are all in it together," he said, stating that the federation was daily confronted "with cases of factories, smelters and production sites shutting down as a result of either skyrocketing operational input costs, or simply by their products being substituted by imports". Citing data from SEIFSA's 2013 *State of the Metals and Engineering Sector Report 2013*, Chibanguza reminded the audience that the M&E sub-sector's output was 20% below what it had been in 2007 and that imports into the country had risen to "more than 50% of domestic demand".

The answer, Chibanguza explained, lay in the sub-sector becoming more internationally competitive, which entailed both an increase in productivity and keeping a lid on production costs.

In my address to the pre-bargaining conference the following day, I expressed concern that it was tempting, during wage negotiations, "for us to forget some very basic truths and to consider ourselves adversaries to one another". The truth was that employers and labour were locked in a symbiotic relationship: business's success is also labour's success because only profitable companies can employ more people, and business's loss was also labour's loss because "failing business ends up letting go of workers and eventually closing down".

"The employer associations represented by SEIFSA have no illusions that the negotiations heralded by this plenary session will be easy," I said. "We understand that they will be tough, with both employers and trade unions defending their respective demands. It is our hope, however, that we will be able to bargain collectively in a mature and constructive manner, with the interests of the country and, in particular, of the metals and engineering sector in mind." In conclusion, I stated that it was clear from the economic diagnosis presented a day earlier that we had a shared understanding of the poor state of our economy, and I argued that it behoved us, collectively, to "guard against worsening the situation – by protecting existing jobs and doing everything possible to create a climate where new jobs can be created".

Sadly, the pre-bargaining conference was a waste of time. With employees' representatives and leaders of employer associations in attendance, it was just an occasion for these parties to play to the gallery. The facts did not matter, and nobody paid them any heed. Labour remained firm on its demand for a 20% increase in wages and a surfeit of other demands, notwithstanding the terrible state of the economy.

However, labour was not the only party that was indignant and, therefore, unrealistic in its expectations. So, too, were employer associations, including their federation SEIFSA – and former President Henk Duys was a co-architect of that approach. It was he, after all, who had advocated – in his personal, "unmandated views" – in the federation's 2013 Annual Report, a move away from what he considered "one-sided bargaining". Employers, he said, were looking for "real benefits" and could only justify increased wage costs if these were offset by productivity gains.

"But before we even start talking about wages and conditions of employment [next year], we must talk about how we are going to deal with the violence and intimidation question, the strike and picket rules and how these are to be enforced," Duys wrote.

Indeed, that was the approach that was to be taken by the SEIFSA Council. For the very first time in its history, SEIFSA would go to the negotiating table armed with a list of its own demands and not only to listen and respond to labour's demands. There were two fundamental demands, from which we were not to budge in the negotiations. First, the SEIFSA Council *demanded* the signing of a peace agreement between employers and labour, as a pre-condition to the commencement of substantive negotiations. Not only would the peace agreement see all parties in the negotiations committing to responsible bargaining, but they would also declare that, in the event of a strike, there would be absolutely no resorting to violence. Labour's refusal to agree to the signing of such an accord would automatically lead to a deadlock. Second, the SEIFSA Council *demanded* the concluding of an agreement on picket rules, which would apply in the event of a deadlock which would lead to strike action.

Even before the negotiations began, the SEIFSA Council also instructed me to write to the Commissioner of the South African Police Service (SAPS) to alert her to the fact that we would soon be conducting wage negotiations in our sub-sector and that there would be the possibility of violence if we reached a deadlock. So, even as it hardened its attitude, the SEIFSA Council knew well that that would lead to industrial action which would be accompanied by violence, hence it wanted us to take pro-active action to counter that possibility.

I wrote to SAPS Commissioner, General Mariah Phiyega, informing her about the looming negotiations and our fears of a possible violent strike and asking for a meeting with her. She and her senior team – including then Gauteng SAPS Provincial Commissioner, General Mothiba – met me at SAPS headquarters on 9 July 2014. They assured me that, upon receipt of my letter, measures had immediately been put in place throughout the country to counter such violence, were it to take place. They asked that some individuals among the companies in the SEIFSA stable volunteer to serve as a link within the police's Stability Unit.

General Mothiba told me that, two weeks into the strike, the police had already opened 49 cases and arrested 103 people and had successfully opposed bail in some of the cases. He stressed the need for businesses to have good relations with police stations in their respective areas. He also expressed regret at the general tendency of business, once a strike was over, for companies to withdraw charges against employees who had been involved in violence.

The demands drawn up by SEIFSA were sent to the MEIBC, in writing under my signature, for distribution to the unions and, bound by the SEIFSA Council's mandate, I duly communicated them when the negotiations eventually began at the Birchwood Conference Centre in Boksburg. The demands came as a big surprise to the unions. There was consternation within the ranks of their leaders, and the NUMSA leadership was seething with anger. Since SEIFSA had never previously entered negotiations with a set of demands, it appeared to NUMSA that it was my doing. It seemed that I had come with an unreasonable approach in an effort to endear myself, as the organisation's first black CEO, to the SEIFSA Council. It appeared that I had said to the white SEIFSA Council: "Leave NUMSA to me. I am black like them. Let me sort them out".

Labour responded with one voice: we are here to negotiate in good faith and are not at war with one another; therefore, we will not sign a peace accord. In fact, labour went as far as accusing us of not taking the process seriously and negotiating in bad faith. They said that it was only once a dispute had been declared in the course of the process that they would be prepared to engage with employers on picket rules.

Ideally, negotiations should be concluded by June, which is the end of the financial year in the sub-sector, so that the increases agreed upon could be implemented with effect from 1 July. Therefore, the more time that passed without any progress being registered in the negotiations because of the SEIFSA Council's demands, the more frustrated everybody became. On SEIFSA's side, we wanted to avoid the possibility of a strike, and NUMSA wanted to secure the best increases possible for its members and to ensure that these were implemented from July, the first month of the new financial year. Since it did not take NEASA seriously and knew that an agreement could be reached without that association coming on board, NUMSA's disappointment and anger were directed at SEIFSA, particularly at me.

To display its strength and seriousness, NUMSA organised a march on the SEIFSA offices at 42 Anderson Street and ensured that its members were there in numbers. They took over Anderson Street, which was closed to traffic, and a big truck carrying the NUMSA leadership parked in front of our building. I was required to accept their memorandum. Operations Director Lucio Trentini, who was familiar with the union's habits, came with me as I walked to the NUMSA van from which a loudspeaker blared. I was asked to climb onto the back of the truck to listen to NUMSA

President Andrew Chirwa reading their memorandum, and then to accept it and say a few words.

Also present was Zwelinzima Vavi, former General Secretary of the Congress of South African Trade Unions (COSATU), who was closely allied to NUMSA at the time. To my surprise, not only was MEIBC General Secretary Thulani Mthiyane out on Anderson Street (his office was opposite our building, Metal Industries House), but he was also on the back of the truck with the NUMSA leadership. I found this shocking because, although he had previously been a NUMSA official, as MEIBC General Secretary, he was supposed to be neutral and to serve both labour and employers. NEASA Chief Executive Officer Gerhard Papenfus could not stand him because of his background, but nobody on the SEIFSA side seemed to mind it.

As Chirwa read the memorandum, my eyes moved from one part of the crowd to another, and I saw anger written on their faces. Since the NUMSA membership had been made to believe that I was the ogre with horns, some choice insults were hurled at me. When he finished reading the memorandum, Chirwa handed it to me and gave SEIFSA a certain period (possibly 10 days) within which to respond to it. Simultaneously, he shoved the microphone into my hand. I was brief: I acknowledged receipt of the memorandum and promised to pass it on to the SEIFSA Council.

During this period, SEIFSA Board and Council meetings took place more frequently. As a rule, the SEIFSA Council met at least once a week to receive feedback on the state of the negotiations and, if they so chose, to revise their mandates accordingly. In the first few weeks of the negotiations, the SEIFSA Council stood firm: we had to sign a peace accord with the unions *before* we could participate in any substantive negotiations with them.

Towards the end of May, no progress had been registered in the negotiations. We would arrive at the bargaining venue at Birchwood to restate our demand for the signing of a peace accord, and labour would dismiss us out of hand and restate its demands, to which we would not respond. That would be the end of the day of negotiations. The cash-strapped MEIBC had paid for the hire of the venue for the whole day, but it was used for no more than two hours a day.

As frustration mounted, so, too, would the vitriol directed at SEIFSA and me continue. At the beginning of the negotiations, part of our mandate was to work with the independent employer associations outside the SEIFSA fold, which were also part of the process, and to try to take them along with us. That included NEASA. Although we differed on the negotiating style and the language used, we all wanted the same thing, according to our mandates: to rein in the historically high wage increases in the sub-sector and to ensure that the increase agreed upon was linked to inflation. It was easy for us to do so collectively at the beginning of the process, but differences eventually emerged and the employers' caucus fractured. In the end, we represented only SEIFSA, while NEASA and the Border Industries Employers' Association spoke for themselves.

The kind of anger caused by the approach mandated by the SEIFSA Council, coupled with the fact that the demand for the signing of a peace accord before substantive negotiations even began had the effect of delaying the beginning of real negotiations, turned the unions against me. Suddenly, I was painted as an arrogant, know-it-all moron who thought he knew better – and publicly described as such. The longer frustration grew with the lack of progress in the negotiations, the more of a target I became. Not only did NUMSA General Secretary Irvin Jim unleash his bulldog, the union's choleric spokesman, Castro Ngobese, on me, but he also went on Radio 702 and personally had a go at me – by name.

In a vulgar opinion piece published by Politicsweb headlined "SEIFSA CEO Kaizer Nyatsumba a rented agent of the white bosses – NUMSA" (which, a decade later, still continues to enjoy pride of place on the website) – Ngobese called me choice names ("a maverick CEO", among them) and held me directly responsible for the stalemate in the talks. "In honesty," he frothed, "Nyatsumba is a supreme example of a rented agent to perpetuate and drive racist, colonial slave wages by White bosses dominating the industry. Nyatsumba must explain why it is only the Black and African working class who are on strike in the metals and engineering industries".

As a by-the-way, it never ceases to amaze me how black people tend to reserve their worst insults for other black people in leadership positions, and rarely take aim at South Africans of other races. Would they have done the same, for instance, if I were Brian Angus or David Carson? Although I always responded to Jim's insults on radio and television, without ever playing the man, I did not bother to do so in the case of Ngobese's drivel. I could not bring myself to stoop to his level. I found that *infra dignitatem* for me. I can only wonder whether, looking back at the situation a decade later, he and Jim are proud of the way they conducted themselves back in 2014.

At a subsequent Board meeting, Chairman Ufikile Khumalo proposed that the Board issues a public statement re-affirming its full confidence in me. Also published in Politicsweb in full, that statement, dated 14 July, reads as follows:

> The Board of the Steel and Engineering Industries Federation of Southern Africa (SEIFSA) continues to have full confidence in the Chief Executive Officer of the Federation, Kaizer Nyatsumba, SEIFSA President Ufikile Khumalo said today.

> Speaking after an urgent Board meeting in Johannesburg this afternoon following a personal attack on Mr Nyatsumba by the leadership of the National Union of Metalworkers of South Africa (NUMSA), Mr Khumalo said that Mr Nyatsumba has done a great job of representing the Federation on all matters, including on the deadlocked wage negotiations that began four months ago.

> Mr Khumalo, who chairs the SEIFSA Board, said that the Board was surprised by the attack launched on the Federation's CEO because everything that Mr Nyatsumba has done in leading the SEIFSA negotiating team and articulating the organisation's views was consistent with his mandate from the Board and the Council.

"We find it regrettable that, at a time when we should all be working together to reach an agreement and to end the strike currently affecting our industry, we have to spend time responding to personal attacks on our CEO. We are happy that, since assuming office in November last year, Mr Nyatsumba has raised SEIFSA's public profile considerably and ensured that the Federation's voice is heard on many matters.

"As a Board, we expect that the engagement between all parties must be responsible and matured", Mr Khumalo said.

It was only once labour had declared a dispute in the negotiations that, worried about the possibility of industrial action, the SEIFSA Council finally dropped its demand for the conclusion of a peace accord before we could engage in substantive negotiations. However, one important condition remained: SEIFSA would not sign an agreement that did not include a clause that would prevent labour from approaching individual companies to open some of the matters for shop-floor bargaining. That followed the striking down, by Judge van Niekerk, a few months earlier of Section 37 of the 2011 Main Agreement of the MEIBC. That court decision seriously concerned companies which were members of the associations affiliated to SEIFSA that they wanted that door closed. Accordingly, SEIFSA wanted labour to commit to a re-written clause that would protect individual companies from shop-floor bargaining subsequent to the conclusion of the industry-wide negotiations.

So, when negotiations resumed, we dropped the insistence on a pre-negotiation peace agreement but reiterated that we wanted such an agreement to be part of the negotiated settlement. However, we stood firm on the need for a mutually acceptable solution to be found to the Section 37 conundrum. It was only then that negotiations began in earnest. Although the MEIBC had engaged the services of an arbitrator, the two sides were poles apart and could not reach an agreement. By then, NEASA had withdrawn from the process.

Section 7 of the Main Agreement states:

(1) Subject to subclause (2)—

(a) the Bargaining Council shall be the sole forum for negotiating matters contained in the Main Agreement;

(b) during the currency of the Agreement, no matter contained in the Agreement may be an issue in dispute for the purposes of a strike or lock-out or any conduct in contemplation of a strike or lock-out;

(c) any provision in a collective agreement binding an employer and employees covered by the Council, other than a collective agreement concluded by the Council, that requires an employer or a trade union to bargain collectively in respect of any matter contained in the Main Agreement, is of no force and effect. (2) Where bargaining arrangements at plant and company level, excluding agreements entered into under the auspices of the Bargaining Council, are in existence, the parties to such arrangements may, by mutual agreement, modify or suspend or terminate such bargaining arrangements in order to comply with subclause (1). In the event of the parties to such arrangements failing to agree to modify or suspend or terminate such arrangements by the date of implementation of the Main Agreement, the wage increases on scheduled rates and not on the actual rates shall be applicable to such employers and employees until the parties to such arrangement agree otherwise.

Once the strike had begun, the Minister of Labour, Mildred Oliphant, got involved in an effort to broker an agreement. We held regular meetings with her in her office in Pretoria and in the business lounge of the hotel at OR Tambo International Airport. The longer the strike continued, the more divisions manifested themselves among the SEIFSA member associations. Ken Manners, of the South African Auto Components Manufacturers Association, urged a speedy resolution of the strike. His company was part of an international supply chain to an original equipment manufacturer in the car manufacturing sector, and he was worried about the consequences of his company's inability to meet its commitments.

Eli Gordon and Kevin Gough of the South African Engineers and Founders Association (SAEFA) were among those who would not budge. In fact, Gough announced at a tense SEIFSA Council meeting that he would be prepared to shut down his company for six months while he went fishing in order to get the NUMSA leadership to "sober up". Some, who spoke tough during SEIFSA Council meetings, were privately among the most concerned. Among them were Henk Duys and Hannes van der Walt, who attended some of our bilateral meetings with the NUMSA leadership.

By then, we had held several such meetings with the NUMSA leadership. Early on during the negotiating process, we hosted Jim and his team for talks over dinner at Country Club Johannesburg in Woodmead. The SEIFSA delegation included members of our negotiating team, some Board members and a selected number of CEOs of some of our member companies. On 13 July, Khumalo and I met with Jim and some of his key lieutenants at his office. In a letter to Khumalo, Jim had fired off *ad hominem* attacks on me, and I seized on that opportunity to express my bitter disappointment at his conduct. Among the concerns that he had raised in his letter was my public communication to update member companies on the state of play in the negotiations. It seemed that Jim wanted NUMSA to be able to communicate at will in the media, but that SEIFSA should not have been doing the same.

"General Secretary", I addressed Jim, "I do not know how you and your office have dealt with SEIFSA in the past. However, please note that I will not be bulldozed or bullied by you or anyone else. I will not be told by you or anyone else what to communicate, when to communicate it, and to whom to communicate it. I will certainly not be told by you what I should communicate to the SEIFSA Council and how to interpret information that becomes available to us. I suggest, sir, that for you to do so is deeply disrespectful towards the office that I hold and the organisation that I represent."

As the SEIFSA Council had feared, much violence accompanied the NUMSA strike. Not only were people prevented from going to work, but trucks were set alight and factories forced to close. Some people who were known to be going to work were attacked at their homes. The longer the strike continued, the worse the mayhem became on the ground. We called on the police to act against those involved in the violence, and the police – understandably – called on employers not only to lay charges, but also to be willing to attend the offenders' court trials once the strike was over and to support the prosecution process by making video footage available. Very few companies were willing to take those steps.

As always happens, in the end – in the fourth week of the strike – we reached a compromise agreement which bore no resemblance to what either side had initially demanded. We reached an agreement on Section 37 of the Main Agreement and a three-year wage settlement – it was only the third time in SEIFSA's history that a three-year agreement had been concluded – which saw low-paid employees getting increases of 10% in 2014, 9.5% in 2015, and 9% in 2016, with higher-paid artisans getting increases of 8% in 2014, 7.5% in 2015, and 7% in 2016. It was agreed that those things which could not be discussed during the negotiations, and which were not part of the new Main Agreement, would be added to the long agenda of the Industrial Policy Forum, which had been agreed upon three years earlier. Jim, Solidarity General Secretary Gideon du Plessis and I signed the agreement at Benoni Lakes Hotel on behalf of our respective constituencies on 29 July 2014.

While I was celebrating the return of industrial peace in the sub-sector, I received a letter from Reserve Bank Governor Gill Marcus asking for a meeting with me for a briefing on the negotiations. The request seemed odd at first, but it was during the meeting itself that Chief Economist Henk Langenhoven and I understood Marcus's interest. With her senior colleagues alongside her, Marcus asked us how SEIFSA settled on 8–10% with labour at a time when the inflation rate was about 6%. Were we not worried, she asked, about the increases' inflationary effect on the economy?

Langenhoven and I explained the context in which the three-year settlement agreement had been reached, including NUMSA's leveraging of violence to achieve its goal. We told Marcus and her team that although it was SEIFSA's original intention to pay only inflation-linked wage increases, the length and effect of the strike on the sub-sector eventually led to the SEIFSA Council settling on those percentages. We also explained that even that settlement had only been possible thanks to the mediation of Minister Mildred Oliphant, who enjoyed the respect of the NUMSA leadership. The spectre of AMCU's five-month-long platinum mining strike had also hung in the air during our negotiations.

## The Still-Born Industry Policy Forum

The Industry Policy Forum (IPF) received its first official mention within SEIFSA from Executive Director David Carson in 2011, in his report to the SEIFSA AGM. In that report, Carson stated that when the negotiation settlement was concluded with labour on 18 July 2011, SEIFSA and all the trade unions "agreed to take immediate steps to establish an Industry Policy Forum to provide senior industry leadership and engage at ministerial level on a number of crucial and urgent challenges facing the metal and engineering industry". This agreement, Carson explains, was considered "a fundamental element" of that year's settlement agreement.

The settlement agreement concluded in 2011 was even worse than the one reached in 2014: increases of between 8% and 10% were awarded for 2011, between 7.5% and 10% for 2012, and between 7% and 10% for 2013 (SAPA, 2014).

The IPF, then, was a joint creation of employers, as represented by SEIFSA, and the labour unions. It was agreed upon after the two parties had spent considerable amounts of time "analysing and examining the significant challenges facing the industry, including the urgent need to create and sustain decent jobs and competitive manufacturing capability in the domestic and global markets". Among the IPF's standing agenda items would be the identification of factors which had contributed to the M&E sub-sector's decline, the development of strategies to secure the long-term interests of employers, large and small, and the viability of regional wage dispensations and exemptions from prevailing agreements; the formulation and implementation of strategies to promote job retention and creation; and the development of strategies on trade and policy matters, designation and export opportunities.

From Carson's (2011) exposition, it is clear that the IPF was a very important creation by the key stakeholders in the M&E sub-sector at the time. Writing in his report as SEIFSA President two years later, Duys (2013) described the IPF as "a mini-industry CODESA" that was agreed upon because industry players could not "continue on the self-destruction path of the past". He said the forum had been expected to bring about "major change and improvement in our industry".

However, the IPF never got off the ground. NEASA, which challenged both the MEIBC's legitimacy and the 2011-2013 settlement agreement in court and won, is alleged to have been the reason for the IPF's failure to get started.

However, in the 2014 and 2017 wage negotiations, it was again agreed that controversial issues which had not been discussed or on which no agreement could be reached were to be set aside for discussion in the IPF. Between 2014 and 2021, the IPF was not convened even once – and again there are various excuses adduced as reasons. One such reason is the subsequent financial instability of the MEIBC, which ended up being placed under supervision by the Labour Court.

To an independent observer, it would appear that while there may have been high hopes for the IPF when it was first agreed upon in 2011, in reality the forum had become a ruse: those tricky or divisive issues that parties to the MEIBC negotiations could not agree upon were relegated to the IPF, never to be heard of again – until the next round of negotiations. In the last five years of my tenure as CEO, the IPF was not talked about at all. It did not feature much even in the 2017 and 2020 wage negotiations.

## Subsequent Negotiations

Given the complaints that I had received in 2014 from some companies that claimed not to know what had transpired in those negotiations and how we had arrived at the percentage increases that we had settled on, I resolved to ensure that there was direct communication with member companies. Traditionally, it was the duty of member associations to liaise closely with their member companies and to seek mandates from them, which they then articulated at meetings of the SEIFSA Council, but whenever I visited the CEOs and Managing Directors of companies, I received complaints they were kept in the dark about the goings-on at the negotiations. All they were told, at the end of the process, was how much they would be required to pay their employees that year.

As a solution, I launched a newsletter called the *2017 Wage Negotiations Update*, which was sent regularly by email to all companies affiliated to SEIFSA member associations and posted on the federation's website. In the inaugural issue, which I launched on 18 April 2017, I encouraged the companies to participate actively in their respective associations' preparations for the wage negotiations and to hold those associations accountable.

"The SEIFSA Office operates strictly in accordance with mandates derived from the SEIFSA Council," I explained. "The latter is an assembly of the 25 Associations that are members of SEIFSA, which brings the member associations together to debate matters among themselves and develop a mandate for the federation. Our responsibility is then to implement the mandate/s coming from our constituency, through the SEIFSA Council."

There was a lot of traffic on the SEIFSA website during the period of the negotiations, which meant that I had accomplished my goal of getting the companies informed throughout the process.

The second issue of the *2017 Wage Negotiations Update* was distributed on 17 May 2017. It reported that the structure and format of the 2017 wage negotiations had begun to take shape, and that we had already received demands from two unions, namely NUMSA and the United Association of South Africa (UASA). The first round of negotiations was set for 7-8 June 2017, and the second one was scheduled for 15 June 2017.

Those negotiations were radically different from the ones that had preceded them. For a start, SEIFSA reverted to its historical stance of not tabling preconditions as demands to be agreed upon before substantive negotiations begin. Second, and much to our relief, NUMSA came to the negotiations with only three demands: a 15% wage increase, a two-year agreement and "finalisation of other outstanding issues". The atmosphere was very different, perhaps thanks to the terrible state in which the economy found itself. The M&E sub-sector was bleeding and a growing number of companies were retrenching employees.

By 6 August, yet another three-year agreement was reached, based on a model that provided wage increases based on actual wages received by employees. The terms of the agreement were as follows: wages would be adjusted by 7% across the board in 2017, by 6.75% in 2018, and by 6.5% in 2019. Again, the IPF can was kicked down the road: it was agreed that "a range of strategic and vitally important industry issues" would continue to be discussed during the currency of the agreement. Jim, Gideon du Plessis and I signed the agreement at a function at the MEIBC offices.

There was not a single day lost to industrial action.

The next round of negotiations was scheduled for April–May 2020. However, the advent of the COVID-19 pandemic hitting our shores in March that year made it impossible for the MEIBC to convene face-to-face negotiations. During the SEIFSA Council's consultative meeting on the approach to take in the negotiations, some individuals argued that the federation should ask for a year-long postponement of the negotiations, with no adjustment to employees' wages that year. This view enjoyed the support of Operations Director Lucio Trentini, who led the negotiations, and the federation's President at the time, former NUMSA negotiator Elias Monage.

I was very uncomfortable with that position. Not only did I consider it to be shortsighted because labour was certain to demand a higher increase when negotiations eventually took place in 2021, but I was concerned that it was not only companies that were badly affected by COVID-19 and the country's lockdown but employees were similarly affected, if not more so. Although we at SEIFSA had forfeited salary increases, I still felt that ordinary employees at these companies needed to have their wages adjusted for inflation – even if that was by a token 3%. At least one of the CEOs participating in the meeting via Microsoft Teams shared my concern, but that view did not win the day.

I was surprised when, after initially pushing back, the labour leadership eventually agreed to a zero-percent salary adjustment for the year. Consequently, employees in the sub-sector earned in 2020 exactly what they were paid in 2019, which made them worse off than they had been a year earlier. As Trentini, Monage and others celebrated this outcome, I considered it a pyrrhic victory.

Indeed, SEIFSA discovered during the 2021 negotiations, when I had already left the organisation, that labour was determined to claw back the increases that had been due to them in 2020. In the negotiations, NUMSA's opening demand was 8%, while SEIFSA's opening offer was 4.4%. A report in industryALL (2023) states that "the union argued that workers sacrificed for the survival of the industries when they agreed to no wage increases in 2020 to mitigate the impact of COVID-19 on the sector". When an agreement could not be reached following NUMSA's declaration of a dispute in the MEIBC, a violent, three-week strike ensued, which cost the industry more than R600 million per day in lost revenue (Trentini, 2024). It was not until 21 October 2021 that a settlement agreement was reached. The percentage increases agreed upon ranged from 5% for artisans to 6% for low-paid employees in the 2021/2022 and 2022/2023 financial years, based on minimum rates of payment. Both parties committed to apply to the Minister of Labour for the agreement's extension to non-members. Labour Minister Thulas Nxesi duly extended that agreement to non-members in the sub-sector with effect from 17 October 2022.

The 2024 negotiations were relatively brief. It took only three rounds of formal negotiations before an agreement was reached between SEIFSA and NUMSA on a three-year settlement, and this occurred well ahead of the expiry of the 2021/2024 agreement. Yet again, outstanding issues were referred to "various working groups and committees for further investigation, discussion and processing" (Trentini, 2024). According to the agreement between the protagonists, wage increases on minimum rates of payment would be 6% for artisans and 7% for unskilled employees in the first year, and then 5% for artisans and 6% for unskilled employees in the second and third years, respectively (Trentini, 2024).

However, Solidarity was not a signatory to that agreement. Instead, the union – long considered a moderate in the industry – slammed the settlement as being "bad news for skilled employees" since it was based on minimum rates of pay, and not on the actual amounts earned by artisans (Du Plessis, 2024).

"This compromise resulted in skilled and experienced employees receiving increases of 3% or less over the three-year period, when the average CPI [consumer price index] was close to 6% for the same period," said Solidarity General Secretary Gideon du Plessis in a press release.

## Conclusion

AMCU's five-month-long strike in the platinum mining sector at the beginning of 2014 set the tone for the approach that NUMSA and other unions adopted in the M&E negotiations of the same year. Seen as the most radical union until then, NUMSA was worried about the effect that a more radical AMCU would have on its membership in the industries to which it might extend its reach. Even then there were fears that AMCU was eyeing the M&E sub-sector with a view to joining the MEIBC as a party to its negotiations.

The SEIFSA Council's ill-advised decision to compile a series of demands, especially its insistence that SEIFSA should not be involved in substantive talks until a peace accord was signed, further poisoned the atmosphere. Occurring as it did in my first few months in office as CEO, this led to the NUMSA leadership wrongly concluding that I was responsible for the federation's combative approach to the negotiations.

Given the then-prevailing state of the economy, the 2017 wage negotiations went smoothly. However, only SEIFSA signed that agreement, on behalf of its affiliated associations, but SAEFA joined NEASA in repudiating that agreement. As a result, it was not extended to non-parties. In the 2021 negotiations, in which NUMSA wanted to claw back the wage increase forgone by its members the year before when it acquiesced with SEIFSA by agreeing to a zero-percent wage adjustment, the union adopted a more hard-line approach, demanding an 8% increase. In the end, the parties settled on 6% for low-paid workers and 5% for artisans. The same thing happened when a three-year settlement was reached in 2024, although that agreement was repudiated by Solidarity.

# Chapter 8

# Moving SEIFSA from a Deficit
# to a Surplus

## Introduction

In April 2014, as we approached the end of the 2013/14 financial year, SEIFSA was forecast to have made a loss of R9.6 million, against a budgeted loss of R8.8 million. Revenue generated in the course of the year was forecast to be R23.6 million, 1.3% above budget, but at R33.1 million, the expenditure was forecast to be 3.4% higher than budgeted. The EC Division was the only one to make its budget for the year, with all the others failing to do so. The KwaZulu-Natal region, where we had a Regional Manager whose responsibilities included revenue generation, was the worst performer, forecast to finish the year 63.2% below budget.

On the expenditure side, by far the biggest concern was rising legal costs, which were forecast to be 180.6% above budget as a result of arbitration between SEIFSA member associations and NEASA on the MEIBC seat allocation. Although the cost was supposed to be covered by companies affiliated to the member associations, not all companies – or, indeed, associations – had paid the levy agreed upon for the resumption of the arbitration process.

The 2014/15 budget was premised on a 7% increase in membership fees, a 4.14% increase in the technological levy income, a 14.3% increase in income generated by the SEIFSA Training Centre and an upward adjustment of the revenue to be generated by the Divisions of SEIFSA, as follows:

- IR (20.6%)
- EC (141%)
- Human Capital and Skills Development (11.9%)
- Safety, Health, Environment and Quality (157.7%)

· KwaZulu-Natal (212.2%), and
· Net income of R1 million from the new Legal Division.

On the expenses side, the budget provided for litigation, the continuing seat-allocation arbitration, more aggressive marketing and public relations, and the registration of SEIFSA in three neighbouring countries. To accomplish the growth envisaged in the three-year strategy approved by the Board, the 2014/2015 budget provided for a net deficit of R12.476 million, based on a 36.11% increase in revenue and a 34.4% increase in expenditure. The forecast was that, in the following year, the federation's net deficit would be reduced to R9.45 million and then to R4.75 million in 2016/2017. A year later (2017/18), SEIFSA would break even and generate a surplus of R46 186.

## Financial Performance during my Tenure as CEO

The targets set for the 2014/15 financial year were deliberately tough. Not only did they require from the CFO and me weekly monitoring of our performance against budget, but they also placed considerable pressure on everybody at SEIFSA, especially the divisions whose duty it was to bring in revenue through their training and consulting activities. Our performance against budget was a standing item on the agenda of Executive Committee meetings. At least quarterly, we would hold budget review sessions to come up with suggestions for new revenue streams that would allow us to make the budget.

We also watched costs very carefully. We did not incur expenditure simply because it was budgeted for, except in the case of the Southern African Metals and Engineering Indaba and the SEIFSA Awards for Excellence, which were introduced during that financial year.

The fact that the financial year began with a month-long strike in July did not help. With companies worried about the industrial action and uncertain about the levels at which a settlement would eventually be reached in the negotiations, there was little demand for SEIFSA's products and services. Therefore, we – like companies in the sub-sector – started the year on the back foot.

The economic climate for the year continued to be tough. Following the conclusion of the 2014 wage negotiations, companies embarked on massive retrenchments, as I had forecast in my speech at the MEIBC pre-bargaining conference in March that year, and some went on to be liquidated. The impact on the sub-sector (and, therefore, on SEIFSA) was significant. Not only was demand for SEIFSA's products and services negatively affected, but so, too, were membership fees for the year.

When it was revised in the course of the year, the 2014/2015 budget provided for a deficit of R5.4 million, with revenue expected to total R39.2 million and total expenditure expected to come in at R42.7 million. By year end, the budget deficit closed at R6.6 million, R1.2 million worse than the budgeted amount. Expenditure peaked at R39 million, thanks to various cost-containment measures. Among the latter efforts, we reduced the huge leave liability by requiring employees to take leave due to them or stand to lose it, with leave due to an employee each year capped at 25 days. The worst performer for the year, with revenue of R324 801, the SHEQ Division was eclipsed even by the new Legal Division, which generated R462 229.

For the 2015/16 financial year, the budget provided for a continuing reduction in the deficit to R4 570 755 million. As it became clear that we would miss our target for the year, it became necessary to take even tougher measures to reduce costs. In addition to a razor-sharp focus on revenue generation, we extended similar attention to costs, which were not declining fast enough. It became clear that we had to revise our initial three-year strategy, which was focused on growth, and focus as sharply on our costs.

Various efforts to tinker with costs had made a difference, but they had not brought us closer to a break-even point. We had to reduce fixed costs by embarking, for the very first time in SEIFSA's history, on retrenchments. At the time, we had a staff complement of 43, inclusive of five positions that were vacant. After following the provisions of Section 189(7) of the

Labour Relations Act, we let go of six employees and eliminated the five vacant positions. Altogether, the process affected eleven positions, and we subsequently re-assigned to remaining employees the work that those who had been retrenched used to do. We ended up with an established staff complement of 32.

We also continued to review the staff leave liability and to rationalise it further. In the 2014/2015 year, we budgeted for a 3% salary increase for everybody, but for the next two years we totally sacrificed salary increases in our quest to end SEIFSA's losses and to get the federation to break even.

We also parted ways with the Deputy CEO, Elsa Venter, who resigned. That led to a huge saving in our fixed costs because I did not replace her. Instead, her responsibilities were divided between Lucio Trentini as Operations Director and me as CEO.

With KwaZulu-Natal Regional Manager Kylie Griffin's services shared with the KwaZulu-Natal Engineering Industries Association, we negotiated with the association to take her over fully as its Executive Director – and made an additional saving.

These were tough actions which had to be taken to save SEIFSA. They required both courage and a great sense of fairness as the Executive Committee met to consider positions that would be superfluous in the redesigned structure. That was relatively easy to do in the case of some of the positions which had been established a year earlier, but it was much tougher in the case of those who had worked for the organisation for years.

Inevitably, the retrenchments affected staff morale and became a major talking point within the bigger SEIFSA family. With Venter having spent her entire working life at SEIFSA and having been well connected with the leaders of our member associations, voices began to be heard at meetings of the SEIFSA Council asking about her departure and about what was going on at SEIFSA. "Have we got the right CEO?" Hannes van der Walt asked at one Board meeting.

By March 2016, the year-end forecast for the 2015/16 financial year was that SEIFSA would finish the year 14% off budget, with the deficit standing at R5 216 781 against a budgeted deficit of R4 570 755. Although not all the divisions made their budgets, with SHEQ again being the worst off (it had made only 39.93% of its target), we had managed to save 22.3% from our operational costs: instead of the budgeted R40 156 446 expenditure, we had spent only R31 198 923, registering a saving of R8 957 523. The initial 2016/17 budget was based on a reduced deficit of R2 650 235, resulting from total revenue of R26 555 546 (the same as in 2015/2016) and total expenditure of R28 404 380. That would represent an improvement of 49.2% in financial performance. Following pushback by the Board, we prepared a break-even budget for the year, which was finally approved.

Although income generated from membership fees decreased by 20.2% because of the resignation from SEIFSA of both SAEFA – until then the largest employer association affiliated to the federation – and the smaller South African Fastener Manufacturers Association, we still exceeded our target for the year. For the first time, we exceeded our revenue budget by 8.1% and, once again, reduced total expenditure by 19% relative to the previous financial year. Therefore, over a three-year period, we had progressively reduced SEIFSA's losses and finally returned the federation to a surplus.

In the 2017/2018 financial year, income generated from membership fees increased by 6.83% from the previous year, and income from the sale of products and services increased by 40.44%, thanks to *SEIFSA At 75*, a profitable magazine that I published that year to commemorate SEIFSA's 75th anniversary, and the success of both the Southern African Metals and Engineering Indaba and the SEIFSA Awards for Excellence. For the first time in three years, we awarded salary increases in the region of 7%, hence total expenditure increased by 6.57%.

These results were illustrated in the CFO's Report in the 2017/2018 Annual Report as indicated in the following graphic.

## KEY FINANCIAL RATIOS

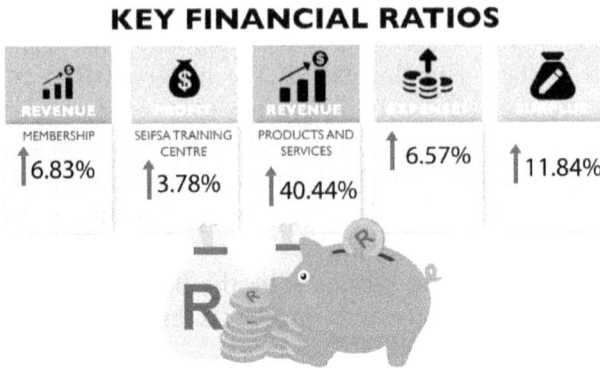

| MEMBERSHIP | SEIFSA TRAINING CENTRE | PRODUCTS AND SERVICES | | |
|---|---|---|---|---|
| ↑6.83% | ↑3.78% | ↑40.44% | ↑6.57% | ↑11.84% |

*Source: SEIFSA Annual Review 2018*

As CFO Rajendra Rajcoomar reported in the 2019 SEIFSA Annual Review, the federation's 2018/2019 budget "was a surplus built on 108% of the prior year's surplus". With a professional culture then fully embedded in the organisation, yet another surplus had been achieved – for the third year in a row. In a year when some companies in the sub-sector consolidated or merged, membership fees increased by 2.83% relative to the previous year, and revenue from the sale of products and services similarly increased by 19.39%. With further salary increases awarded to a deserving team that continued to work hard, expenditure for the year increased by 7.19%.

The results were illustrated in Rajcoomar's report in the 2019 SEIFSA Annual Review as shown in the following graphic.

## KEY FINANCIAL RATIOS

| | |
|---|---|
| Revenue: Membership | ↑ 2.83% |
| SEIFSA Training Centre Profit | ↑ 14.38% |
| Revenue: Products & Services | ↑ 19.39% |
| Expenses | ↑ 7.19% |
| Surplus | ↑ 24.44% |

*Rajendra Rajcoomar*
Chief Financial Officer

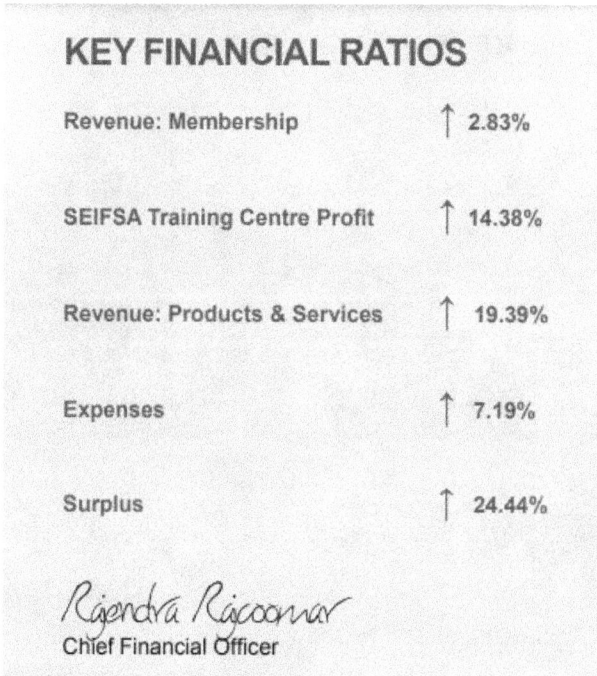

*Source: "SEIFSA Annual Review" 2019*

Thanks to the black-swan event called COVID-19, which led to a lockdown in the country, SEIFSA's by-then-established positive trend of great performance was brought to a halt. Although we were ahead of budget in the first six months of the financial year, our performance in the last four months of the year (March to June) was badly affected. In the climate of that pandemic, companies were understandably very reluctant to send employees to attend training courses or to have people from outside come in for consultations. Although we had again budgeted for a surplus, we recorded a loss at the end of the 2019/2020 financial year.

Compared to the previous year, membership revenue decreased by 4.72%, the SEIFSA Training Centre generated no profit whatsoever, revenue from products and services declined by 16.84%, and expenditure declined by 3%.

Rajcoomar captured the 2020 financial performance in the 2020 Annual Report as indicated in the following graphic.

## KEY FINANCIAL RATIOS

| | |
|---|---|
| Revenue: Membership | ↓ 4.72% |
| SEIFSA Training Centre Profit | ↓ 100% |
| Revenue: Products & Services | ↓ 16.84% |
| Expenses | ↓ 3.00% |
| Net result | ↓ 13 times last year's performance |

*Rajendra Rajcoomar*
*Chief Financial Officer*

*Source: SEIFSA Annual Review 2020*

That was to be the last set of results for which I was responsible as SEIFSA CEO.

## Conclusion

Historically, SEIFSA has been heavily dependent on membership fees for its survival. When the economy became tough in the early 2000s and companies began to lay off employees, the federation – whose membership fees are based *per capita* on employees – felt the pain in the pocket. Together with the majority union in the sub-sector, NUMSA, SEIFSA submitted a proposal to the Minister of Labour, as provided for in the country's labour dispensation, for those companies and employees that were unaffiliated to pay a CBL that would benefit the two organisations.

That CBL lasted for ten years, and SEIFSA harvested more than R55 million from it. The money came in very handy, making it possible for the federation to continue to focus heavily on collective bargaining and IR. During his tenure as Executive Director, Brian Angus introduced some training and consulting services, to be accessed by member companies for a fee, to supplement SEIFSA's revenue.

The expiry of the CBL on 31 December 2012 affected SEIFSA badly. After many years of being financially sustainable, having built considerable reserves, SEIFSA registered a loss at the end of the 2012/13 financial year. For the 2013/14 financial year, the Board approved – for the very first time – a deficit budget in the region of R8 million. That meant that SEIFSA began to dip into its reserves for its continued survival.

Following my appointment as CEO on 1 November 2013, I implemented numerous changes at the federation, getting it to function as a professional business enterprise, and implemented a jointly-developed three-year turnaround strategy which saw the organisation breaking even and generating surpluses. Having broken even, SEIFSA generated profits in 2017, 2018 and 2019 and, like all other corporate entities, it was terribly affected by COVID-19 and the lockdown that accompanied it in 2020.

Predictably, these achievements were not without challenges of their own. The changes that I introduced at SEIFSA to make these results possible led to acrimony and much nasty corporate politics, as will be clear in Chapter 11.

# Chapter 9

# Introducing Innovations at SEIFSA

## Introduction

Although its fortunes have waxed and waned over the years, the manufacturing sector in South Africa has been an important part of the economy for decades. Not only has it been a significant contributor to the country's GDP over this time, but it has also offered employment to many. Within manufacturing, the M&E sub-sector has been of strategic importance because of its role in infrastructure development.

Yet, for many years, South Africa did not have a forum at which those with an interest in manufacturing – or, indeed, in its M&E sub-sector – could congregate to deliberate on its health. However, mining conferences there were aplenty. I was told that SEIFSA held a one-day SEIFSA Conference annually, at which matters of common interest to affiliated employer associations and member companies were discussed, but I do not remember seeing any programme. Instead, when I was told in 2014 that the response to that year's SEIFSA Conference was poor, I saw an opportunity to launch a larger, industry-wide conference for the entire sub-sector, beyond affiliation to SEIFSA. Taking place over two days, the professionally organised Southern African Metals and Engineering Indaba (for that was its name) would be open to all interested stakeholders, regardless of their attitude to SEIFSA. It would be an annual M&E conference, and SEIFSA would merely be its organiser.

Early in 2015, Liz Hart launched the Manufacturing Indaba, an event which has since become entrenched in the annual life of the sector. The M&E Indaba was launched at the same venue, Emperor's Palace, on 28–29 May 2015.

My biggest concern was the obvious disconnect among key stakeholders, in particular business, labour and Government. It was clear to me that it would take a joint search for solutions by the three stakeholder groups to revive the M&E sub-sector. That is why I said the following at the beginning of the M&E Indaba: "We believe firmly that business alone does not have all the answers, that labour alone does not have all the answers, and that the Government alone does not have all the answers. Instead, we believe that it is vital for the three stakeholder groups to work very closely together, in a solid partnership, to advance the interests of South Africa and the Southern African Development Community region."

## Resistance to Change

There was general resistance to change within the SEIFSA community. The M&E Indaba was heavily marketed within SEIFSA, beginning with the affiliated associations and their member companies. While there was excitement in some quarters, the usual nay-sayers within the SEIFSA Council were not pleased with the idea, merely because it had not been done before. They were such creatures of habit who were reluctant to give new ideas a chance. With the Board's support, I ploughed on.

I was clear that the first year of the conference was going to be an investment. The idea was to build the M&E Indaba so that it became an established and profitable conference in subsequent years, but the best-case scenario in its first year was for it to break even. I had incentivised staff members to raise sponsorships for the conference, with 10% of the value of the sponsorship to be paid to them as commission. In the conference's first year, the platinum sponsor, to the value of R300 000, was the MEIBC, while the Department of Trade and Industry was the primary value-in-kind sponsor and endorser.

Although he had accepted the invitation to be one of the speakers at the conference, Trade and Industry Minister, Dr Rob Davies was subsequently said to be unwell and was represented by Deputy Director-General for Industrial Policy, Garth Strachan. The whole SEIFSA Board descended on Emperor's Palace, where they rubbed shoulders with Government and labour leaders and participated enthusiastically. Also present were labour leaders

Irvin Jim and Gideon du Plessis, of NUMSA and Solidarity, respectively. For the session focusing on Eskom, the panel included a member of the Official Opposition in Parliament, the Democratic Alliance.

Further attempts to keep the Department of Trade and Industry on board were unsuccessful. On 25 February 2016, Angela Dick – who was SEIFSA President at the time – invited me to a meeting with Strachan in Pretoria. At that meeting, Strachan complained that at the inaugural M&E Indaba the previous year he had found himself "whipped from all sides the whole day". He said that, in future, it would be important for us to frame the relationship "to establish how do we work with SEIFSA in a collegial way".

The following image depicts an advertisement for the 2017 M&E Indaba as carried in the *Sunday Times*.

**DON'T MISS OUT!**

The Third Southern African Metals and Engineering Indaba This Week, 14 - 15 September 2017, Sandton, Johannesburg.

14 - 15 September 2017

High-profile Speakers Include:

| ANC Treasurer-General | Former African Union Chairperson | ANC MP | NUMSA General Secretary | MD: Emerging Markets Africa: Deloitte |
| --- | --- | --- | --- | --- |
| Dr Zweli Mkhize | Dr Nkosazana Dlamini-Zuma | Dr Makhosi Khoza | Irvin Jim | Dr Martyn Davies |

Other prominemt speakers include:

- Bheki Ntshalintshali - General Secretary: COSATU
- Nico Vermeulen – Director: NAAMSA
- Gideon du Plessis – Solidarity General Secretary
- Bonang Mohale - CEO: Business Leadership South Africa
- Lindiwe Zulu - Minister: Small Business Development
- Mildred Oliphant - Minister: Labour
- Charl Folscher - Managing Director: Voith Turbo
- Tanya Cohen - CEO: Business Unity South Africa (BUSA)
- and Many More Quality Speakers.

**FOR BOOKINGS, CONTACT:**
Nuraan Alli: 011 298-9436
or nuraan@seifsa.co.za

**SPONSORSHIP OPPORTUNITY:**
Contact, **Melissa Sibindi:**
011 298-9456
or melissa@seifsa.co.za

**ONLINE BOOKINGS:**
www.meindaba.co.za

Don't Miss Out. Book Now on www.meindaba.co.za.

*Source: SEIFSA 2017*

Clearly, Strachan had found it uncomfortable to be asked tough questions by business and labour leaders who wanted to know how the Government planned to help the struggling M&E sub-sector. His colleague, a lady by the name of Desiree, revealed that Minister Davies was not ill at the time of the inaugural M&E Indaba, but he had chosen not to attend the conference because he was concerned about "the Government's share of voice". Davies, it seemed, would have preferred a conference where he and his officials would have done all the talking, without the involvement of other stakeholders. To my surprise, Angela Dick added that she had also been "uncomfortable with the presence of the opposition"!

The conference was a great success. Attended by 280 delegates, the Indaba had seven sponsors and 20 exhibitors and generated a lot of news coverage. Although it made a loss in its inaugural year, in subsequent years we had it hosted at the Industrial Development Corporation Conference Centre in Sandton, in partnership with that company, and it generated handsome profits each year. The profits generated from the M&E Indaba contributed directly to SEIFSA breaking even and then generating surpluses three years in a row.

Over the next few years, speakers at the M&E Indaba included Trade, Industry and Competition Minister Ebrahim Patel (pictured below), Local Government and Co-operative Governance Minister Dr Nkosazana Dlamini-Zuma, then Finance Minister Pravin Gordhan, and former Leader of the Official Opposition, Mmusi Maimane.

*Source: SEIFSA Annual Review 2020*

The leaders of NUMSA and Solidarity have been a regular feature at the conference, as have fraternal organisations like the National Association of Automotive Manufacturers of South Africa and the National Association of Automotive Component and Allied Manufacturers. Other key participants have included academics from some of the country's top universities.

## The SEIFSA Awards for Excellence

As was evident in an earlier chapter, there is great concern in the M&E sub-sector about the influx into the South African market of cheaper imports from south-east Asia, which seriously threaten those involved in manufacturing. Some companies that were previously involved in manufacturing have since become traders, merely importing goods into the local market and selling them at a mark-up to make some profit. The dominant view in the M&E sub-sector is that the answer lies in the imposition of protective tariffs and a more rigorous enforcement of localisation or designation of products that are manufactured locally.

While import tariffs and designation have a role to play, in the longer term the answer must lie in innovation and investing in technologically efficient plants to reduce the unit cost of production. The answer, therefore, lies in South African manufacturers being internationally competitive so that they can hold their own in the domestic and international markets.

That is why I introduced the SEIFSA Awards for Excellence in 2015. I wanted to encourage innovation in the M&E sub-sector to improve the chances of companies being internationally competitive. Just like the M&E Indaba, the SEIFSA Awards for Excellence are open to all companies in the sub-sector, including those that are not members of associations affiliated to SEIFSA. Indeed, for three years in a row, the most innovative company award went to a company which is a member of NEASA.

Some of the trophies won by companies in the respective categories are displayed in the following image.

*Source: SEIFSA Annual Review 2018*

An internal advertisement for the SEIFSA Awards for Excellence, as it appeared in one of SEIFSA's Annual Reports, is shown in the following image.

*Source: SEIFSA Annual Review 2018*

The awards have grown phenomenally, attracting more entrants each year. The following were among the founding categories: Most Innovative Company of the Year, Best Customer Service Award, Environmental Stewardship, Corporate Social Responsibility Award, Most Transformed Company of the Year and Best Artisan of the Year Award. The entries for each category are assessed by at least three judges, with the chairperson of each category being independent of SEIFSA. In the awards' inaugural year, former SEIFSA Executive Director Brian Angus accepted my invitation to be one of the judges and chaired the category in which he was an adjudicator.

## Commemorative SEIFSA Publications

SEIFSA had just turned 70 years old when I joined it in November 2013. That important milestone was allowed to pass uncelebrated. However, when I came on board, we commissioned the *Financial Mail* to publish a high-quality supplement to commemorate the federation's 70th year. The final product, titled "SEIFSA: 70 Years On", was published as an insert in that magazine on 28 March 2014.

*Financial Mail* journalist David van Biljon spent time at SEIFSA interviewing the Executives and other people who were made available to him. These included my predecessors Brian Angus and David Carson, as well as former SEIFSA President Henk Duys. Serving President Ufikile Khumalo and I, who were newbies at the time, merited a joint, short write-up on the supplement's last page.

The cover of the supplement is shown in the following image.

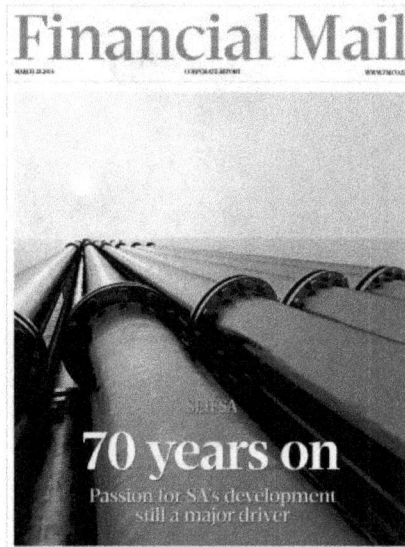

Source: *Financial Mail*, April 2014

A very professionally produced publication, that supplement lay prominently displayed on the table in my office over the years. It was published at no cost to SEIFSA, thanks to the advertising that was generated from companies and others wishing SEIFSA a happy 70th birthday.

I had more time for us to prepare for the federation's 75th anniversary. The federation had a Communications Manager at the time and we planned the publication of a glossy, high-quality magazine that would be a coffee-table book celebrating the anniversary. Communications Manager Ollie Madlala conducted a series of interviews and wrote all the articles, while the Sales, Marketing and Communications team sold advertising space. Yet again, staff members were incentivised to sell more adverts by means a commission of 10% of the value of the advertisement sold.

The front cover of *SEIFSA at 75* is shown in the following image.

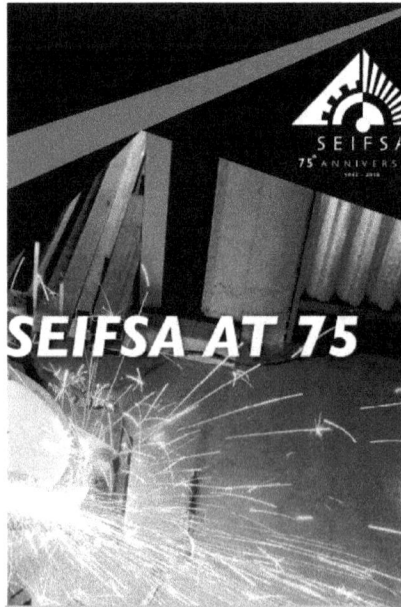

*Source: SEIFSA Annual Review 2018*

Titled *SEIFSA At 75*, the publication was a run-away success. Not only was it visually attractively, but it also generated a profit of just under R1 million!

## Conclusion

As CEO, I had the privilege of introducing several innovations at SEIFSA; these collectively constitute my legacy at the organisation. I am immensely proud of them all, especially since I introduced them in an environment that was not known to be pro-change or in favour of innovation. For all the innovations mentioned in this chapter and elsewhere in the book, I must acknowledge the support and enthusiastic cooperation of then Marketing and Communications Executive, Adelia Pimentel, who was such a pleasure to work with.

I know that some of these innovations – such as the SEIFSA Awards for Excellence – are still part of SEIFSA and continue to grow from strength to strength. I am pleased to have introduced them.

# Chapter 10

# Changes in SEIFSA Leadership over the Years

## Introduction

When I joined SEIFSA in November 2013, a significant percentage of the people in the federation's employ had been there for many years, with some not having known another employer in their lives. That is why they were so comfortable at SEIFSA and hoped to continue to cruise comfortably towards retirement.

The federation's employees and the leaders of the different employer associations affiliated to the federation had struck up close personal relationships over the years and were very comfortable with one another. Some had even become good friends. As a matter of routine, SEIFSA staff members attended the monthly meetings of the affiliated associations and briefed those meetings on their respective portfolios or fielded questions, and they also attended meetings of the organisation's "Executive Committee" and the SEIFSA Council.

The relationship was a symbiotic one and, in some instances, it can even be said to have been nepotistic and incestuous. They protected one another; that was how close they were. An environment like that is not the most welcoming to strangers.

That is why, as changes occurred within the SEIFSA team, some individuals within the affiliated associations became ultra-critical of me as CEO. They held me responsible for the departure of some employees, even when they had moved on to better – or better-paying – opportunities elsewhere, as was often the case.

In this chapter, I document some of the more important staff changes during my tenure as SEIFSA CEO.

## Staff Movements

The first person to leave SEIFSA following my appointment was a tall, frail man whose name I remember vaguely as Joe Pitman. He had reached SEIFSA's retirement age of 63, but he had been given a three-month extension as a special dispensation. I was told by Deputy CEO Elsa Venter, shortly after I had come on board, that Pitman's initial three-month extension was almost over and that he had asked for another extension of the same duration. I duly approved the new extension, but asked that he be informed that I would not agree to another extension at the end of the three-month period. As that period approached expiry, I stood firm and did not approve another extension. A young lady, Neo Zulu, was subsequently employed in the position he had been in.

Human Capital and Skills Development Executive Nazrene Mannie was recruited by Angela Dick to join Transman. Although she left just as I was settling in, I appreciated her help in getting Mustak Alli appointed as her successor.

Next to leave was IR Officer Brett Carson, who I was told was a University of the Witwatersrand graduate. He left to join a company to which he used to consult in the sub-sector. Over the next few years, it was to be the norm for us to lose employees to member companies which were familiar with and impressed by the work of SEIFSA staff. I understood that it would not have been easy for him to continue to work at SEIFSA when he knew my strong views on nepotism.

Finance Executive Nicolene van Huyssteen needed a change of scene for health reasons. I asked Finance Manager Baakile Motha, who was approaching the end of her maternity leave, to return to work earlier as Acting Finance Executive. For three months Van Huyssteen was Executive Assistant to the CEO, but eventually felt that she had to step away from work altogether to focus on her recuperation. The CFO position eventually went to Rajendra Rajcoomar.

Later, when Motha resigned to take a position at a company closer to her home in the north of Johannesburg, I turned down Rajcoomar's plan to replace her with a South African of Indian descent. It was my philosophy that all divisions had to be fully diverse and that none should be dominated by people from a certain ethnic group, that is, that the person who headed the division in question should not appoint people of the same racial or ethnic identity as their own. When he presented me with the second-placed Mariaan de Jager as his choice, I eagerly acceded to her appointment as Finance Manager – and what a wonderful appointment that turned out to be.

Brian Angus's son, Gordon Angus, left in 2015 to work with Ross Williams and Kevin Gogh as Executive Director of the South African Engineers and Founders Association (SAEFA). He went from being Operations Director Lucio Trentini's protégé to being his and SEIFSA's opponent and chief critic. When SAEFA later resigned from SEIFSA as a result of ideological non-alignment, we had a series of meetings with the CEOs and Managing Directors of companies in that association (many of whom knew only that their companies were members of SEIFSA) and retained most of them in other SEIFSA member associations. When Gordon Angus left, I combined his position with Mokoetle's and appointed her IR and Legal Services Executive, again realising a saving.

I have already explained elsewhere in the book that Elsa Venter resigned. I abolished the Deputy CEO position, and we made a huge saving.

For me, the biggest losses were those of Marketing and Communications Executive Adelia Pimentel and Chief Economist Henk Langenhoven in 2016. I was happy for Pimentel when she was appointed Executive Director of the Perfumes Association of South Africa, but I was sad to lose her. She was a hard-working and conscientious team member who fully supported my vision from day one. Langenhoven, who was close to retirement, was given a five-year contract as Chief Economist of the Chamber of Mines. That would see him remaining gainfully employed past SEIFSA's retirement age, so I was happy for him, but sad to see him go. He was a consummate professional.

I wanted to replace Langenhoven with Tafadzwa Chibanguza, but I felt that he needed to spend some time in the role to grow into it. Therefore, I appointed him Senior Economist, to run the EC Division, but told him of my intention to give him about a year in the role before elevating him to the Chief Economist position. He had finished his Master of Commerce degree by then. Sadly, Langenhoven came back for Chibanguza and offered him a job in his team at the Chamber of Mines. I then appointed Dr Michael Ade as Chief Economist.

Buoyed by its bigger budgets, the Chamber of Mines (now called the Minerals Council) came calling again, this time for Human Capital and Skills Development Executive Mustak Alli. I then asked for a meeting with my counterpart there, Roger Baxter, and asked him to refrain from raiding SEIFSA staff members. To his credit, he promised to stop the practice – and we did not lose anybody else to the Chamber of Mines.

Although most of these staff movements made sense as the incumbents took better-paying jobs, some members of the SEIFSA Council – like Williams, Gogh, Louis Breckenridge and Peter Flint – repeatedly railed at me, questioning "the quality of the SEIFSA Executive Leadership".

After he had spent three years with us, Dr Ade moved on to an undisclosed entity, and I then appointed former ArcelorMittal South Africa Chief Economist Chifipa Mhango as our Chief Economist. Both Ade and Mhango are naturalised South Africans.

Nonhlalo Mphofu, the SHEQ Executive, headed a division for which there seemed to be little demand. Year after year, the SHEQ Division did not make its budget, which meant that it was a financial burden and did not add value. It made sense to abolish the position and to offer SHEQ services through contracted third parties. Accordingly, Mphofu was retrenched, but SEIFSA ended up offering even more SHEQ services through joint-venture partnerships.

Perhaps the toughest resignation for me to make peace with was that of Mokoetle. For some time, she had shared with me her intention of returning to legal practice, and for a while I had asked her to put it off. Eventually, though, she felt strongly enough that the time was right for her to open her own legal firm. She worked for us on a contractual basis for about three months while her successor, Sibusiso Mthenjana, was settling in.

I had worked very well with Mokoetle, who was a very principled colleague with strong points of view. She was not the most popular member of the Executive Committee because of her strong character, and yet I found her to be the most reliable colleague whose judgement I trusted implicitly. She was alive to the virulence of the corporate politics within the SEIFSA community and often expressed concern that there were moves afoot to eject me from the organisation. Each time she raised this concern, I asked her not to worry because the conspirators would not have anything to use against me.

Mokoetle was also an effective Company Secretary who tried her best to ensure that the Board observed good corporate governance. When she felt it was necessary, she wrote tough memoranda to Board members, holding them to SEIFSA's MoI and the Board Charter. I missed her sorely in the last eighteen months in my job at SEIFSA, but both Trentini and Rajcoomar were happy to see her back.

The following image shows SEIFSA Executive Committee at the end of the 2018 financial year, with the Management Team depicted below that.

EXECUTIVE TEAM

Kaizer Nyatsumba
Chief Executive Officer

Lucio Trentini
Operations Director

Melanie Mulholland
Human Capital and Skills Development Executive

Michael Ade
Chief Economist

Bridgette Mokoetle
Industrial Relations and Legal Services Executive

Rajendra Rajcoomar
Chief Financial Officer

Nonhlalo Mphofu
Safety, Health Environment and Quality Executive

MANAGEMENT TEAM

Mariaan de Jager
Finance Manager

Michelle Norris
Human Capital & Skills Development Manager

Mark Lotter
Marketing Manager

Marique Kruger
Economist

Theresa Crowley
Associations Manager

Nuraan Alli
Sales Manager

Michael Lavender
Industrial Relations Manager

Zolile Moyikwa
IR & Legal Services Manager

*Source: SEIFSA Annual Review 2018*

## Conclusion

During my years as CEO of SEIFSA, several staff movements took place. Most of these movements were voluntary, with employees being offered bigger or better-paying jobs. The federation, which was going through a stringent belt-tightening exercise to ensure that it broke even after years of suffering losses, did not have the means to counter the offers that were made to those staff members. Certainly, SEIFSA could not compete with the financial muscle of the Chamber of Mines, which for a while targeted SEIFSA for recruitment.

I am happy for all the staff members who gained opportunities to grow in different environments, but their loss was acutely felt at SEIFSA. Sadly, in an environment in which the perceptions of the leaders of some employer associations were influenced by prejudice, these staff movements offered potent ammunition to those opposed to the changes that I introduced as CEO. The last time this concern was raised, during a February 2021 SEIFSA Council Meeting, Alpheus Ngapo, the SEIFSA President at the time, responded: "Talent is the most mobile asset".

# Chapter 11

# The Continuing Struggle with Good Corporate Governance

## Introduction

Ross Williams must have been elected Chairman of SAEFA, SEIFSA's largest member association at the time, in the course of 2014. I know that I first met him months after the 2014 wage negotiations, when he told me, in the course of a visit to his office on the West Rand, that he wanted to see a different outcome to the next round of negotiations.

Until then, I had known Eli Gordon and Kevin Gogh from that association, both of whom had been very outspoken in meetings of the SEIFSA Council. While Gogh generally seemed to be of a choleric disposition, Gordon seemed to have taken a strong dislike to me from the very beginning. Even before the expiry of my six months' probation, he was already going around from one Board member to another to campaign for me not to be confirmed in the job. When they were later joined by Williams on the SEIFSA Council, they made a very vocal triumvirate that often lobbed unfounded criticism at me disguised as criticism of "the SEIFSA Executive Team".

They were often assisted by Peter Flint of the Light Engineering Industries Association and Constructional Engineering Association (CEA) Executive Director Louis Breckenridge. Although his job was to serve the CEA by representing its interests with relevant stakeholders, including SEIFSA, Breckenridge often considered his job description to be to pass his and his Executive Committee's requests on to SEIFSA for us to carry them out as instructions. A garrulous man with a loud voice, he made sure that he had his tuppence's worth on every issue, even after the Chairman and Vice-Chairman of his association had already made their respective contributions on those issues. I am not aware of a single instance of him going out himself to implement his Executive Committee's decisions or instructions.

At his last SEIFSA Council Meeting on 1 August 2016, Flint launched a full-frontal attack on me, without once naming me. He had a major go at "the quality of the SEIFSA leadership," without adducing any shred of evidence to support his allegation. Board Chairperson Angela Dick just sat there, saying not a word. Flint stopped short of calling for my dismissal, and only Atlantis Foundries CEO and SEIFSA Vice-President Pieter du Plessis dared to object to the unsubstantiated criticism.

During that meeting, Flint stated that SEIFSA was poorly led and that NEASA CEO Gerhard Papenfus offered "real leadership" in the M&E sub-sector. At the time, there was no doubt in my mind that both my race and the transformation agenda I unashamedly stood for were the issues that deeply concerned the likes of Gordon, Gogh and Flint, and which triggered their antipathy towards me. Two days later, Ross Williams wrote to me to offer an explanation on behalf of Flint: the latter, he said, had informed him that his intention was "to confront the issues about the executive that seem to be constantly bubbling under ... in other words, to either get behind the executive and support them fully, or to replace them".

There was that lack of logic again. It confirmed again that some members of SEIFSA's affiliated associations wrongly considered themselves – and not the SEIFSA Board – to be the bosses of the SEIFSA CEO and his Executive Team. Some continued to labour under this mistaken impression even when they had the federation's MoI in their possession.

So strongly did I feel about the insult heaped on me (under the guise of "the SEIFSA leadership") by Flint *et al.* that I directed written communication to the Board and asked for an honest discussion of what was going on. I asked that, at its next meeting, the Board should be bold enough to discuss everything, "including matters that some people may find uncomfortable to discuss or engage with openly, frankly and robustly". Never before, throughout my working career, had I felt singled out for vitriol based on my race and what I stood for.

In response, Williams accused me of "playing the race card". In a longer submission ahead of a Special Board Meeting, I took him head on:

> While it is not acceptable for anybody to try to silence either debate or criticism by "playing the race card", in Ross Williams's words, regrettably there are some among our white compatriots today whose actions and pronouncements continue to be informed, either consciously or unconsciously, by racism. Over the years some of them have grown so arrogant as to silence any black person who may have a legitimate reason to point their racial prejudice out or to complain about such conduct as "playing the race card". For some unfathomable reason, it is as if they can say anything they want about black people, motivated by their conscious or unconscious racial prejudice, but black compatriots have no right to call them out.
>
> In my humble opinion, that is hypocrisy of the worst order, which is itself symptomatic of latent racism. There is reason to believe that some of these white compatriots would never dare accuse a Jewish person who complains about conduct that s/he regards as antisemitic of "playing the semitic or Jewish card", but they make bold as to hurl such accusations churlishly at black South Africans who complain about conduct that they consider racist.
>
> I am deeply offended by the gross insensitivity, exhibited in Ross's callous comment in his letter of 3 August 2016 to me, to the experience of black South Africans during the apartheid years. The truth is that one has to have experienced something like racism to be able to perceive or detect it when it rears its ugly head. It is not for somebody who was never on the receiving end of such a pernicious system to tell its past victims that they are "playing the race card"!

Racism against black people existed in South Africa for centuries and was codified in the noxious form of apartheid – which the United Nations correctly described as a crime against humanity – for almost 50 years. A change of laws, following the adoption of our Interim Constitution in 1994 and the new Constitution in 1996, was a very important start towards the eradication of that scourge, but it did not change anti-black prejudices that were deeply ingrained for many years in the conscious of some of our white compatriots. This is a fact that we, the Board, and all South Africans will do well to accept.

But Williams, Gordon, and company were not the only challenges that I encountered during my tenure as SEIFSA CEO. By far the most daunting challenge, which ultimately led to my premature resignation, was dealing with some of the federation's Presidents, in their capacities as Board Chairpersons. During those periods, good corporate governance was observed only in the breach, with Chairpersons acting as though they were entitled to obeisance from their fellow Directors.

## Board Chairpersons for Less than a Term

When I joined SEIFSA in November 2013, all members of the Board – then erroneously called the "Executive Committee" – were elected to office for just one year. That meant that, at the next AGM, a theoretical possibility existed that all of them could be replaced by new members of the governance structure. Not only was there no stability in the "Executive Committee", but the governance structure's members did not have much institutional knowledge of the organisation and had to be educated on the same matters each year.

As a Certified Director at the time (I have since become a Chartered Director [SA]), I knew that stability and institutional knowledge were important in a governance structure. I could not countenance having to deal each year with different Board members who had only superficial knowledge of the matters which mattered most in the life of the federation. Therefore, among the first amendments that I proposed to SEIFSA's MoI in

February 2014 were the following: that Board members be elected for a two-year term of office and that each year the term of half of the directors should expire. Upon expiry of their terms, they would be eligible for re-election, should they so wish. Both the Board and the AGM approved these amendments.

The President's term of office remained one year, but to be eligible for election as President, a Director would have to have served at least a year on the Board. However, upon expiry of his term (until then, all of them had been men), a President could stand again for re-election. Although throughout SEIFSA's history some men had served as President more than twice, none had served more than two years in a row in that position. Care had always been taken to ensure that the incumbent did not repeatedly occupy the position.

There were two important aspects of the MoI that remained unchanged: the federation would continue to have three Vice-Presidents. The First Vice-President would succeed the President at the next AGM, and the out-going President automatically became the Second Vice-President. Very rarely, therefore, was the Presidency contested. Theoretically, that would happen only if the First Vice-President passed away or resigned during the year, but neither scenario occurred during my tenure at SEIFSA.

During my eight years at SEIFSA (strictly speaking, seven years and nine months), I worked under seven Presidents or Board Chairpersons. When I joined the organisation in November 2013, Henk Duys – who had been the President the year before – was the Interim President, pending Ufikile Khumalo's assumption of office. Khumalo's first Board meeting was in February 2014. At the AGM later that year, he was elected to serve another year as President, meaning that he was SEIFSA President during his two years on the Board.

A senior executive at the Industrial Development Corporation who had been appointed Chairman of Scaw Metals, Khumalo was by far the best Chairman of the federation's Board during my era. He was fair and unflappable and always presided over Board and SEIFSA Council meetings with equanimity. He was a man of integrity who did not need to throw his weight around;

he could always be counted upon to be resolute and firm when the situation demanded.

While Khumalo was serving his second year in office, Aveng Trident Steel Managing Director Alpheus Ngapo was the first Vice-President and was due to succeed Khumalo as President. As I drove to The Campus in Bryanston in the morning of the AGM, I received a call from Ngapo, who informed me that, owing to some differences that he had with the Aveng Group, he would be withdrawing his candidature for the presidency. My attempt to get him to reconsider his decision failed. The next two most senior Non-Executive Directors were Hannes van der Walt and ABB CEO Leon Viljoen; their terms were ending at that AGM and they were not standing for re-election. Instead, the Chairpersons of SAEFA and the CEA, Williams and Ben Gerrard, respectively, were up for election for the first time.

I feared that, as chairpersons of their respective associations, Williams and Gerrard were unlikely to owe their primary loyalty to the federation, as was required of all Board members. Instead, while they were on the board, there was a high probability that they would consider themselves activists for their respective associations, notwithstanding the injunction against such activity. As Company Secretary, Bridgette Mokoetle had sent to all Board nominees the General Principles which guided elected members' conduct. The seventh principle states: "In keeping with Good Corporate Governance, once elected, all SEIFSA Board Members will have a fiduciary duty to the federation, respect the Board Charter and serve only its interests, and not those of their respective companies or associations".

Angela Dick, founder and CEO of the labour broking company Transman, was going into the second year of her term, so I turned to her, out of frustration, and asked if she would be happy to stand for election as President. Not only was she going to be the most senior corporate leader on that Board, but I also thought that her election would ensure that SEIFSA continued to make history: Khumalo was the first black person to be President, and he would then be succeeded by the first woman ever to occupy the position.

Within days of Dick's election, it became obvious that I had made a big mistake by turning to her during a period of crisis. It soon became clear that, although she had been part of a team ably led by Khumalo in the first year of her tenure, she saw the SEIFSA presidency as yet another executive role, to hell with the organisation's MoI and the Delegation of Authority approved before she joined the Board. Suddenly, she asked me for a copy of my CV and copies of my qualifications, as if I was applying for a new job. She followed that up by informing me that she would be taking over our Boardroom for the next three days, during which time she would like to have one-on-one meetings with me and all members of the Executive Committee. I duly had the Boardroom prepared for her occupation.

Indeed, she spent three full days in our office, with each member of my Executive Committee going in the Boardroom for long periods at a time for scheduled meetings with her. Not only did she ask them about their respective jobs in the federation, but she also peppered them with questions about me and my leadership. Some were asked to prepare written submissions in response to the bizarre questions that were asked. Finally, it was my turn. By then, she had already gone through my CV, all certified copies of my qualifications and my 2013 letter of appointment.

The meeting with me was not as long as those with the members of my Executive Committee had been. She couched it as an opportunity for her to understand the state of affairs within SEIFSA – more than a year after she was first elected to the Board, and after she had spent that year contributing or asking very little during Board meetings.

Next, Dick volunteered the services of her company's CFO and Information Technology Manager, who were assigned to comb through our finances and our information technology system. When they arrived, Rajendra Rajcoomar, who was responsible for both functions in the business, made them feel welcome and provided them with whatever information that they required. In the weeks and months that followed, silence followed the new Chairperson's strange behaviour.

It was clear, though, that she had a particular goal in mind. What it was became clear some months later.

Before that, though, let me introduce to the reader Michael Pimstein, former CEO of Macsteel Holdings (Pty) Ltd, and Elias Monage, former NUMSA chief negotiator in the MEIBC. Due to assume the SEIFSA presidency in October 2013, Pimstein had parted ways with the company, while Monage had crossed the floor and gone into business in the M&E sector. When I met them, the former was Co-CEO of a financial investment company that he had co-founded with friends and the latter was a BEE partner at Steloy Casting, where he was a Board member. Below, I describe how I met them.

One day in 2015, Lucio Trentini told me that Pimstein had called him and asked for a meeting. One of them (I can't remember who) wanted me to be part of the meeting. Trentini said Pimstein had proposed breakfast, and I suggested that we meet at Country Club Johannesburg in Woodmead one morning. In preparation for the meeting, I needed more information about the man, and Trentini duly obliged.

At the meeting, I was more of an observer than a participant. Pimstein asked about different people within the SEIFSA establishment, including former Deputy CEO Elsa Venter, whom he knew well over the years. Trentini explained that Venter had since moved on. After briefly explaining his parting of ways with the proprietors of Macsteel Holdings (Pty) Ltd, Pimstein expressed his desire to return to the SEIFSA fold, particularly the SEIFSA Board. The one option open to him was for him to take up associate membership of SEIFSA, which would entitle him to all the benefits of membership available to those who worked for companies that were members of associations affiliated to SEIFSA.

In terms of the federation's MoI, associate membership was initially approved by the Executive Committee and then ratified by the Board. Once Pimstein's associate membership application had gone through both stages, he could stand for election to the Board or be co-opted to it. Once he was on the Board, he played a very important role as a Non-Executive Director, together with

Neil Penson, Oupa Komane and Bob Stone, in relation to matters of governance.

Dick's presidency was the most challenging and haphazard – and I am being very circumspect here to avoid disclosing confidential Board matters. In May 2016, when we presented a budget for the 2016/17 financial year, it was rejected on the grounds that the associations wanted to be involved in determining SEIFSA's strategic trajectory because some were concerned that we were focusing excessively on revenue generation at the expense of advocacy and lobbying. One individual who pined for the bygone era even suggested that SEIFSA should return to the structure that had existed in 2002! Ironically, some of the individuals who voiced this criticism also wished that their membership entitled them to SEIFSA's products and services, at no cost to them. As things stood, membership of SEIFSA entitled them to being represented in collective bargaining and to advocacy on their behalf, in addition to a significant discount on the cost of the products and services offered by SEIFSA.

Consequently, we presented two strategic options with different cost implications to the SEIFSA Council: SEIFSA could return to focusing strictly on collective bargaining and advocacy and lobbying, or it could continue to offer revenue-generating products and services in addition to collective bargaining and advocacy work. The former option would see membership fees – which were subsidised by the revenue independently generated by SEIFSA – rising considerably, while the latter option would be a continuation of the status quo, even as we undertook to ensure that more attention was paid to advocacy and lobbying going forward.

There was no appetite for higher membership fees at a time when companies were already struggling. Therefore, unanimously, through the SEIFSA Council, the member associations chose the second option, which we termed the hybrid model. This is how we phrased it:

SEIFSA positions itself as a professional collective bargaining entity in the areas of socio-economic and human capital policy and a professional and effective advocacy and lobbying body. This option secures the income derived from membership dues and products and services (i.e. PIPS, STC, and MAH, etc.). A secondary focus is placed on providing professional advisory, consultancy, and training services. Income lost from this stream may need to be subsidised by the membership to hold onto staff members integral to achieving the primary objective.

In a 22-page submission, we then presented a revised budget – which did not differ much from the original one – for the 2016/2017 financial year. Only then was it approved by the Board.

In Board meetings, Dick and Williams, in particular, often saw things the same way. The year of her presidency was one of complete non-alignment not only between the Chairperson and the CEO, but also between the Chairperson and the rest of the Board. In the end, she resigned from the Board a few months before the AGM. Hers was a totally unlamented departure. Shortly thereafter, she resigned from the CEA, whose labour-broking division she had chaired, and joined Williams's SAEFA, which soon announced its resignation from SEIFSA.

At the AGM in October 2016, Pimstein was elected President, with Atlantis Foundries CEO Pieter du Plessis and Alpheus Ngapo (then a senior executive at ArcelorMittal South Africa) as Vice-Presidents. Pimstein had been one of the voices of reason during the Dick presidency and continued to be supportive of me and the executive team I led. Strangely, he asked me to have SEIFSA business cards printed for him, and I duly did so, although I could not fathom why he needed them since he was a Non-Executive Director. As President, he suggested two individuals that he wanted co-opted to the Board: these were Bukelwa Bulo, a member of his company's Board of Directors, and Monage. They were both co-opted to the Board, but Bulo attended just one meeting and then resigned. Another person to be co-opted onto the Board was Mayleen Kyster, Managing Director of RSC Avelo Steel, where she was a BEE partner.

We worked very well with Pimstein throughout much of the year: he spent two full days with me attending the Southern African Metals and Engineering Indaba, and he came with me to some meetings with NUMSA General Secretary Irvin Jim during the 2017 wage negotiations. His meeting place of choice for us was Tashas Restaurant, either at Melrose Arch or in Hyde Park, for breakfast.

Things started going awry in October that year, following a complaint by Nonhlalo Mphofu, in her other capacity as the Manager responsible for the Small Business Hub's (SBH's) activities. A few months earlier, SEIFSA – through the SBH – had won a tender to handle Voith Turbo's enterprise and supplier development (ESD) programme. Among the deliverables was that SEIFSA had to identify small companies that would benefit from Voith Turbo's ESD budget, offer them the required training, and ensure that they were able to meet their respective performance targets. The federation handled the full ESD budget and accounted for it to Voith Turbo accordingly, providing all accompanying invoices and receipts.

Mphofu brought to my attention a concern about the invoicing of Voith Turbo for work done for one of the company's suppliers being developed. I called CFO Rajendra Rajcoomar into the meeting, and he gave an explanation that I found to be satisfactory. At a subsequent, weekly SBH meeting at which Mphofu, Rajcoomar and Operations Director Lucio Trentini were also present, Rajcoomar proffered the same explanation – which was satisfactory to everybody present. However, on 20 October 2017, Mphofu wrote to me, copying in President Michael Pimstein (among others), in the context of a legitimate but unrelated concern that I had raised with her, asking me to appoint an independent third party to review that invoicing. I wrote back to her on the same day and instructed IR and Legal Services Executive Bridgette Mokoetle to engage the services of an independent service provider to conduct the requested review or investigation.

That afternoon, Pimstein wrote to Mokoetle, copying me in, not only directing her to use the services of Moore Stephens, but also instructing that he should be kept informed of "the process details and meetings". Even more bizarrely for a Non-Executive Chairman, he expressed a wish to attend the meetings with the service provider and to be kept copied "in [on] all ongoing communications". I found that both strange and unlike the Pimstein I had come to know and respect, but I kept my peace. It seemed that he had long been made aware of the allegations made by Mphofu. In her reply that evening, Mokoetle reminded Pimstein that Moore Stephens were SEIFSA's auditors. Instead, she proposed using another service provider "to ensure that the process is transparent and independent (and also to eliminate a potential further complaint on the independence of the investigation)". Therefore, she explained, she had already approached law firm Edward Nathan Sonnenbergs (ENS) Africa's forensic audit division for assistance.

Pimstein remained unsatisfied. He went further and instructed Mokoetle, in her capacity as IR and Legal Services Executive, about how to do her job. He instructed her to ask ENS Africa, Moore Stephens, Bowmans, and Howarth, in that order, "to each forward their proposal and timetable together with anticipated cost". As CEO, I had left the matter to the discretion of Ms Mokoetle as the responsible Executive, but here was the Board Chairman directing her to his preferred service providers and seeking to micro-manage the investigation. Then he added ominously: "For good order and given the participants in this matter, any instruction given must be to report to the President/ Chaiman".

I was deeply concerned by Pimstein's conduct and its implications. It created the impression that he trusted only himself, rather than Mokoetle or the CEO, to get to the bottom of Mphofu's concerns. By encroaching on the CEO's territory, he communicated a message that he had no confidence in the incumbents. The following day, on a Saturday, 21 October 2017, I wrote to Pimstein to raise my concerns: "This is an operational matter, and the CEO is dealing with it and should continue to do so to the end while keeping the Board informed. This matter will form part of the CEO's Report for the November Board Meeting. You may recall that that is how other matters, including the serious allegations against a former senior member of staff, were handled."

Pimstein seemed to have been surprised by my views. He shot back an immediate response, which contained a veiled threat:

> I still haven't seen any response to my request for the brief sent to ENS on this matter and it is now evident from Bridgette's recent mail to Ms Mphofu that you are proceeding without recognition of my request. Should you continue to do so and should you continue to confuse the Board's proper functioning with a misdirected understanding of your responsibility and authority, I will have no option but to call an urgent meeting of the Board and take such action necessary to address SEIFSA's interests. Under no circumstances should you ignore my communications in this regard.

That was yet another bizarre response from Pimstein. Since being copied in on Mphofu's concern and on my request to Mokoetle, he had not once spoken or written to me. Instead, he had directed communication to Mokoetle, copying me in, and yet somehow he could stoop to accusing me – and not Mokoetle – of showing no recognition of his request, and he was threatening me with an urgent Board meeting that would supposedly sort me out for I know not what. In response, I assured him that I respected the authority of the Chairman of the Board, that I had never ignored any communication from him or anyone else, and that I had not had any dealings with Mokoetle since asking her to engage the services of an independent third party to investigate Mphofu's concerns.

On Sunday, 22 October 2017, Pimstein wrote to Mokoetle and me, suggesting that we meet over breakfast at Melrose Arch the following morning "to discuss the issues ventilated over the last few days". Mokoetle replied to say that she could not make that meeting owing to prior business commitments, but would be able to make it on Friday, 27 October 2017. When we finally met at Tashas that Friday morning, it was agreed that the investigation by ENS Africa should go ahead as planned, with the findings to be shared with the Chairman and the Board. I duly reported to the Board at its meeting on 6 November 2017 on the investigation, and the Board required me to share the report with it as soon as I received it.

I received the ENS Africa Forensics report on Monday, 13 November 2017, read it and shared it with Trentini and Mokoetle, and then forwarded it to the Board with my interpretation of it and my decision on the matter. Two days later, I received from Mokoetle, in her capacity as Company Secretary, notice of a Special Board Meeting convened by the Chairman, to take place on 29 November 2017 "to discuss the ENS Forensics Report as well as the role of the Board and its relationship with Senior Executives and employees of SEIFSA". I confirmed my availability for the meeting and duly had it diarised. In preparation for that meeting, I asked the Company Secretary to circulate to the Board SEIFSA's Board Charter and the Delegation of Authority Framework, and she sent it through on the same day. I added two submissions on the agenda of that Special Board Meeting: a report on the Business Unity South Africa Integrity Pledge and an update on the 2017-2020 MEIBC Consolidated Main Agreement.

On Tuesday, 21 November 2017, I wrote to Pimstein, copying in all other Board Members, tendering my apologies that I would no longer be able to attend the Special Board Meeting scheduled for 29 November 2017 because "gastroenterologist Dr C. Kassieniedes, to whom my wife was referred by our family GP, Dr Colin Kahanovitz, today booked my wife for admission at 6am at the Morningside Medi-Clinic on Wednesday, 29 November for a gastroscopy and colonoscopy". I added that during my four years with SEIFSA, "I have never once missed a Board or SEIFSA Council Meeting, whether scheduled or special – until now". I received neither a reply to that email nor an acknowledgement of receipt, with only one Board member writing back to me to wish us well with the two procedures.

Concerned about the fact that there appeared to be serious misalignment between Pimstein and me on this matter, and worried about the possibility of a rift between us, I thought it important for us to meet before going on holiday in December so that we could discuss issues constructively and resolve whatever misunderstanding or non-alignment existed between us. I duly reached out to him – via text messages, a voice message left on his cellular phone and an email – to ask that we meet over breakfast or lunch. Pimstein was very busy at the time and was not immediately accessible. When he eventually got back to me via text message, he suggested that we should arrange to meet in January. Regrettably, we did not manage to meet before the Board Meeting of 5 February 2018.

To my utter dismay, that Board Meeting, whose contents I am not at liberty to disclose, was reduced to a circus by Pimstein. In the process, he lost the respect of his fellow Board members. In what he insisted be a dialogue between himself and me, with the other Board members reduced to being spectators, Pimstein accused me of having "cancelled the Special Board Meeting of 29 November 2017". The strange thing was that he knew that, as CEO, I had absolutely no authority to cancel a scheduled Board Meeting – and that I had made absolutely no attempt to do so. Instead, I had merely tendered my own apology; there was no reason the meeting could not continue without me as I had made a written submission, the Board had the ENS Africa Forensics findings and Mokoetle, who had commissioned ENS Africa to conduct the investigation, was going to be present at the meeting.

Based on its investigation, the ENS Africa Forensics team concluded that it "could not conclude that his [Rajcoomar's] actions amounted to fraud or corruption". Instead, he was guilty only of having purchased software for the computer of a Voith Turbo ESD beneficiary without my prior knowledge and approval as CEO, an error that he had acknowledged to the ENS Africa Forensics team. Had the finding been otherwise, I would not have hesitated to subject Rajcoomar to a disciplinary hearing. Upon consultation with SEIFSA's IR guru, Trentini, and our lawyer, Mokoetle, I decided that no disciplinary action was necessary. However, I did require that he pay back the money that had been used to buy the software for the ESD beneficiary, and we implemented the ENS Africa Forensics team's recommendations relating to the formulation and introduction of clear guidelines, policies and procedures governing the SBH and all related projects.

Shortly afterwards, Pimstein resigned as SEIFSA President and Board Chairman. Vice-President Pieter du Plessis stepped in as Interim President.

At that time, Alpheus Ngapo was back on the Board and was again the First Vice-President. He was due to take over as the next President at the October 2018 AGM. However, he and Elias Monage asked for a meeting with me before nominations for Board appointments closed and told me that they had agreed between the two of them that Monage – who was not one of the Vice-Presidents – would run for President, with Ngapo doing so a year later. While I knew Ngapo well at the time, I knew very little about Monage, other than his union background. While the former was a Chartered Accountant with a Master of Business Leadership degree from the University of South Africa, the latter's qualifications were vaguely listed as "a degree in Insolvency Law and Practice" from Rand Afrikaans University, that he had studied Telecommunication, Regulation Policy and Management at the University of the Witwatersrand "and holds a qualification in Executive Coaching from the University of California in Berkeley". The federation did not verify Board members' claimed academic qualifications. In addition to having been a member of the Board of Steloy Casting when he was first co-opted onto SEIFSA's Board of Directors, Monage also described himself as Executive Chairman of Arabela Holdings (Pty) Ltd and Afika Holdings (Pty) Ltd, neither of which had an official website. Since I first met him more than four years ago, he has described himself in his profile as "currently reading for his Master's degree in engineering business management", but the name of the institution where he is studying for this unusual qualification was not given.

In subsequent years, Steloy Casting was first placed in business rescue and then eventually liquidated.

Indeed, at the AGM later that year, Monage was duly elected SEIFSA President, with Ngapo as First Vice-President and Auto Industrial Group CEO Andrea Moz as Second Vice-President. Kyster was also a nominee for that Board. To his credit, Monage revealed at the AGM that she had not covered herself with glory during her first tenure on the Board, been suspected of having routinely leaked sensitive Board communication to Mphofu when the latter was retrenched and subsequently claimed constructive dismissal.

## Conclusion

Embedding good corporate governance at SEIFSA, starting with the introduction of a Board of Directors instead of the ill-named "Executive Committee", was hard work. To ensure that those elected to the Board of Directors knew what was expected of them, Company Secretary Bridgette Mokoetle and I arranged annual corporate governance training sessions that took place in November each year, immediately after the AGM. To avoid any conflict of interest resulting from my being a long-standing member of the Institute of Directors in South Africa, each year we engaged the services of Terry Booysens of CGF Research Institute (Pty) Ltd to conduct training on good corporate governance.

Notwithstanding these efforts to improve the calibre of those elected to the SEIFSA Board of Directors and to acquaint them with the Companies Act 71 of 2008 and the King IV Code on Good Corporate Governance, the SEIFSA Board experienced challenges of different kinds each year. Two of the federation's Board Chairpersons, who lost the confidence of their fellow Directors because of the way they conducted themselves along the way, did not even finish their respective year-long tenures in office.

Regrettably, they were replaced by a new breed of Non-Executive Directors, some of whom used their membership of the Board to advance their own interests rather than those of the federation and its legitimate stakeholders.

# Chapter 12

# Allied Institutions: Non-Existent Corporate Governance but, Financially, a Ray of Sunshine

## Introduction

Neither the story of the M&E industries nor that of SEIFSA is complete without reference to important allied institutions, the aforementioned Metal and Engineering Industries Bargaining Council (MEIBC) and the Metal Industry Benefit Funds Administrators (MIBFA). The federation had a direct hand not only in the conception and eventual birth of both organisations but has also played a crucial role in their continued existence.

In this chapter, we wend our way through both organisations for the sake of completeness. We look first at the constitutional quagmire that the MEIBC has found itself in and then conclude with a discussion of MIBFA's legendary corporate governance failure that was only recently resolved.

## The Metal and Engineering Industries Bargaining Council

Given that SEIFSA was formed to represent the collective voice of employer associations in the M&E sub-sector, one of its very first co-creations – barely a year into its own existence – was a body that would serve as a platform from which the federation would engage with organised labour in collective bargaining. That was the National Industrial Council for the Iron, Steel, Engineering and Metallurgical Industries, which came into existence in 1944 as a byproduct of an agreement between SEIFSA, then still known as the South African Federation of Engineering and Metallurgical Associations, and the trade unions in the sub-sector.

Following the promulgation of the Labour Relations Act in 1995, which legally established bargaining councils, the National Industrial Council for the Iron, Steel, Engineering and Metallurgical Industries registered as a bargaining council, thus becoming a statutory body. A year later, it formally changed its name to become the Metal and Engineering Industries Bargaining Council (MEIBC). The federation was the sole representative of employers in the council until 15 March 2005 when the National Association of Employers of South Africa (NEASA) was admitted as another employer party.

Like all other bargaining councils, by law the MEIBC has four primary responsibilities. These are collective bargaining, compliance management, dispute resolution, and social protection. Its scope covers all aspects of the diverse M&E sub-sector, including general engineering, manufacturing engineering, the lift-engineering industry, and the plastics industry. In terms of the MEIBC's constitution, only "registered employers' organisations and registered trade unions may claim membership". While there is no minimum number of companies stipulated for an employers' organisation to become a member, it is clearly stated that trade unions applying for membership should have at least 5 000 members in the sub-sector (MEIBC, 2011).

In terms of the Labour Relations Act, all employers active in the M&E sub-sector are required to register with the MEIBC. Like all its counterparts, the MEIBC is funded through the payment of agreed levies by companies and their employees. These include the Registration and Administration Levy (RAL), the Dispute Resolution Levy (DRL) and the Collective Bargaining Levy (CBL). When it expired on 31 December 2012, the CBL was not renewed. The RAL and the DRL expired together at the end of March 2015, marking the beginning of a challenging era for the institution because of the MEIBC Management Committee's late referral to the Minister of Labour of a request for the gazetting of the renewal of the 2011 MEIBC Administration and Expenses Agreement. According to NEASA CEO Gerhard Papenfus (2016), part of the problem was the MEIBC's failure to submit its audited financial statements to the Department of Labour because its Management

Committee had refused to accept the statements following "MEIBC officials' unauthorised redirecting of funds".

During what SEIFSA referred to as "a hiatus period", neither employers nor employees were legally obliged to make payments to the MEIBC for the two levies. In its communication to companies in associations affiliated to it, SEIFSA advised employers to refrain from making payments towards both sets of levies until such time as the MEIBC had confirmed the gazetting of the new agreements by the Minister of Labour (SEIFSA, 2015).

What followed was a period of severe financial difficulty for the MEIBC, which could not perform some of its core functions. Attempts to broker a compromise failed when wrangling continued between the employer bodies, primarily SEIFSA and NEASA, with the latter demanding amendment to the MEIBC Constitution to provide that requests to the Minister of Labour for extension to the whole sub-sector of collective bargaining agreements should take place only when there is a 75% majority agreement among the combined employer parties. The situation's deterioration led Solidarity Deputy General Secretary Marius Croucamp to call on the Labour Minister, Mildred Oliphant, to intervene (Kilian, 2016).

"The reality is that the MEIBC will have to finally dissolve within three months if the parties in question cannot agree on the expansion and increase of the Administration and Dispute Resolution levy agreements at the next Management Committee meeting," Croucamp said (Kilian, 2016).

At an urgent MEIBC Management Committee meeting in December 2016, which was attended by senior officials from the Department of Labour, the employer associations outside the SEIFSA stable argued that the MEIBC was insolvent and needed to be placed in voluntary liquidation. On the labour side, Solidarity agreed with the view expressed by NEASA, the Plastics Converters Association of South Africa, the South African Engineers and Founders Association and the Consolidated Employers' Organisation. However, Solidarity parted ways with these organisations when it came to the proposed solution: instead of voluntary liquidation, Solidarity recommended business rescue.

No agreement was reached at that meeting, but at least the Sick Pay Fund was approved for another five-year period.

The following year, Solidarity approached the Labour Court, asking it to appoint an Administrator, with powers similar to those of a business-rescue practitioner, to run the affairs of the MEIBC to ensure that it did not fold. The court granted the order, appointing Afzul Soobader as Administrator for an initial period of six months, with the possibility of an extension. He enjoyed all the powers of a business rescue practitioner and was charged with the responsibility of returning the MEIBC to a state in which it was no longer financially distressed. The federation welcomed the court's decision, describing it as "a victory towards saving the MEIBC" (SEIFSA, 2017).

It took Soobader's intervention – which was extended until February 2020 – in his capacity as an Administrator imbued with the powers of a business rescue practitioner, to get the relevant levy agreements renewed and submitted for gazetting for another five years. SEIFSA was to concede that having the MEIBC placed in administration had saved it from "entering full-blown liquidation and closure" (SEIFSA, 2017). While responsible for the MEIBC, Soobader presented a turnaround plan and budget and oversaw the gazetting and extension to all employers and employees in the sub-sector of the Administration and Dispute Resolution Levy Agreements, which collectively account for up to 95% of the Council's revenue. When these levies expired in October 2020, Labour Minister Thulas Nxesi had them gazetted anew.

## Metal Industries Benefit Funds Administrators

Without doubt, by far the most successful creation of the partnership between SEIFSA member associations and organised labour in the M&E sub-sector is the Metal Industries Benefit Funds Administrators (MIBFA), which was established in 1957. Through MIBFA, the sub-sector has had the industry's collective funds, jointly contributed by employers and employees to the benefit of the latter's social security, managed and invested. Over the years, the following funds were established: the Engineering Industries Pension Fund, the Metal Industries Pension Fund, the

Metal and Engineering Industries Permanent Disability Scheme, and the Metal and Engineering Industries Bargaining Sick Pay Fund.

As a company, MIBFA provides administration services to the aforementioned benefit funds. The Metal Industries Pension Fund (MIPF) and the Engineering Industries Pension Fund (EIPF) are run by a joint Principal Officer who is accountable to Boards of Trustees comprising an equal number of employer and labour representatives. Collectively, the two pension funds are the largest private fund in South Africa, jointly managing R120 billion by October 2020. The MIBFA itself is run by a Board of Directors, also drawn equally from the ranks of employers and labour. Over the years, SEIFSA has provided the employer representatives to that Board and to the two pension funds' Boards of Trustees, most of them drawn from nominations from its member companies.

Within the SEIFSA fold, the MIBFA has generally been acknowledged to have been a run-away success as an entity. Trustees of the two pension funds are regularly provided with the relevant corporate governance and funds training. However, things have been very different for the MIBFA Board of Directors itself, which has been as good as non-existent for many years.

Upon joining SEIFSA in November 2013, I took a seat on the MIBFA Board of Directors and on each of the pension funds' Boards of Trustees, alongside Deputy CEO Elsa Venter and others. Shortly thereafter, I stood down from the pension funds' Boards of Trustees to focus on my fiduciary duty as an MIBFA Non-Executive Director. To prepare for the effective execution of my duties in this capacity, I asked for a copy of MIBFA's MoI and the Board Charter, but I was never furnished with any. Of a Board of six (three employer representatives and three labour representatives), I was the only voice to make this demand. When I continued to insist on these documents, CEO Jacques Calomiti promised to come to see me in my office for a briefing. When he did, he brought with him a tattered, basic document, the Articles of Association, which had been compiled in terms of the moribund Companies Act of 1973 and which did not spell out the duties and powers of the Board of Directors.

For the first four years, Macsteel CEO Hannes van der Walt was Chairman of the Board and NUMSA's Vuyo Mabho was the Deputy Chairman. Board meetings took place twice a year, and each meeting lasted just 30 to 45 minutes. We would arrive at the Boardroom where the meetings were held, spend a few minutes exchanging pleasantries and having refreshments, get started with the meeting (the agenda of which would have four to five items), routinely provide the required approvals (without any debate or discussion), finish the meeting, and leave. A razor-thin Board pack would have been provided two to three days before the meeting, and the sum total of our duties was to approve the budget at the beginning of the financial year and to sign off on the annual financial statements at the end of the year.

The CEO took the scantiest minutes during Board meetings and these were provided in the form of a one-line summary per item, with the minutes often being a total of one or two pages long.

Not once, from November 2013 to May 2020, did the MIBFA Board of Directors discuss and approve any policies or strategies or pose any questions to the CEO, nor was there any corporate governance training session arranged for the Board. The CEO ran the show as he saw fit, he was totally unaccountable to the Board (which was blissfully ignorant of anything to do with the company), and the Board members were content to twiddle their thumbs. Far from being a Board of Directors, we were a rubber-stamp – and I was most uncomfortable with the situation.

It was a most unsatisfactory situation, one in which the unionists and the employer representatives were merely ultra-polite to one another and to Calomiti and his executive team. As a Certified Director at the time, I knew that the situation was unacceptable. After numerous attempts at raising my concerns, both during Board meetings and in private meetings with Calomiti, I resolved – during the COVID-19 lockdown of 2020 – to take the bull by the horn by placing a written submission before the Board, apprising the Directors of our collective dereliction of duty and the concomitant legal liabilities, and demanding an immediate remedying of the situation.

In a memorandum dated 4 May 2020 and headed "Urgent need to ensure that MIBFA Board of Directors complies with the Companies Act to avoid being declared Delinquent Directors", I pointed out the shortcomings of the Board and the way it functioned and demanded that swift action be taken to remedy the shortcomings. In particular, I demanded that:

- We commission a law firm to draft an MoI and a Board Charter for MIBFA, both of which would guide the Board in performance of its duties;
- We appoint an individual or company to provide Company Secretarial services to the MIBFA Board of Directors on a part-time basis;
- We approve a Delegation of Authority Framework for the CEO;
- We set up and populate two statutory committees, the Audit and Risk Committee and the Social and Ethics Committee, each with its own Terms of Reference and a Chairperson;
- We ensure that a minimum of four Board Meetings take place per annum, one each quarter, and receive reports from Management in accordance with an Agenda to be determined by the Board of Directors, as represented by its Chairperson;
- We ensure that the aforementioned Board Committees each meets at least twice a year; and
- We ensure that Non-Executive Directors are paid approved Board fees for attendance at board and Committee Meetings.

"Gentlemen", I concluded (all six of us were men, with Mabho being the Acting Chairman at the time), "being a Director is a serious legal responsibility, which has attendant legal risks. In the event of a company (in this case, MIBFA) falling foul of the law, Non-Executive Directors can be held legally – and personally – liable, be sentenced to prison terms or be declared Delinquent Directors (as is likely to happen soon to former SAA Chairperson Dudu Myeni). We have a legal duty, therefore, to take our responsibilities to MIBFA and all its stakeholders seriously and to hold Management accountable".

Thankfully, the memorandum had the desired effect. At our next Board meeting later that month, the MIBFA Board of Directors approved all my proposals and, in the course of a three-month period, the company had a new MoI and a Board Charter, a Delegation of Authority Framework for the CEO, and Audit and Social and Ethics Committees. I appointed IR and Legal Services Manager Vuyiswa Miya the third employer representative on that Board, with the other one being Operations Director Lucio Trentini. As a Board, we officially elected Mabho as Chairman, following my nomination, and I was appointed Deputy Chairman. When I left SEIFSA a few months later, we were in the process of embedding good corporate governance at MIBFA, which was still in the process of learning how to draft policies that were to be presented to the Board of Directors for its approval. This is one of the legacies of which I am immensely proud.

## Conclusion

Over the years, SEIFSA, working in partnership with unions in the M&E sub-sector, set up a number of worthwhile institutions that have served South Africa well over the years. Among the most prominent were the MEIBC (and its forerunner, the National Industrial Council for the Iron, Steel, Engineering and Metallurgical Industries) and the MIBFA. Like SEIFSA itself, these institutions were not perfect, but they have made major contributions to the country's economic life over the years.

In this chapter, I have focused on the financial and governance challenges that confronted the MEIBC and MIBFA during my tenure as SEIFSA CEO and have indicated how they were resolved. It is my hope that in years to come these institutions will benefit from the resolution of the challenges that they experienced during the 2013–2021 period.

# Chapter 13

# Integrity Takes a Back Seat

## Introduction

Shortly after his election as the new SEIFSA President in October 2018, Elias Monage, who had not first served an apprentice period as Vice-President, asked me to get business cards printed for him. That was the second time I had received such a request from a newly elected SEIFSA President, so I had the Communications Manager order the cards for him.

Also elected to that Board of Directors, for the first time, was Tumi Tsehlo, then CEO of the State-owned South African Mint, who was to go on to form a formidable partnership with Monage. A few months later, Tsehlo moved to the BEE space and became CEO of Dynamic Fluid Control. As was to become clear later, from then on, as BEE entrepreneurs, his and Monage's interests were more closely aligned.

Alpheus Ngapo was the First Vice-President, with the other Vice-Presidents being Andrea Moz and Neil Penson. Also new to the Board, like Tsehlo, were Patrick Metswi of Murray & Roberts, Seneca Lutchmana of Lixil Africa and Nonhlanhla Ngwenya, a BEE partner and Board member at KSB Pumps & Valves (Pty) Ltd. A group photograph of the Board members elected at the 2018 AGM is presented below.

Newly elected SEIFSA Board

*Source: SEIFSA Annual Review 2019*

## Different Rules for Different People

Even before his election as SEIFSA President, Monage – who was one of five members of the South African Chapter of the BRICS Business Council (BBC) – had argued that it was important for SEIFSA not only to be active within the BBC's Manufacturing Working Group (MWG), but also to chair it. He had previously raised this matter with me, but I was lukewarm about it because I had a lot on my plate at the time. He subsequently made the proposal at a Board meeting on 4 February 2019, and the Board was persuaded of the merit of his suggestion. He then undertook to use his influence within the BBC to make this happen.

An image of the letter Monage wrote to the BRICS Business Council on 7 March 2019 to nominate me is presented below.

S E I F S A

Steel and Engineering Industries Federation of Southern Africa
OUR PASSION, YOUR SUCCESS

7 March 2019

Business Constituency
BRICS Business Council

Dear Sir/Madam

**NOMINATION OF SEIFSA CEO KAIZER M. NYATSUMBA FOR THE BRICS
BUSINESS COUNCIL MANUFACTURING WORK STREAM**

The metals and engineering (M&E) sector is a very important part of manufacturing. It
represents around 45% of manufacturing in the country. The Steel and Engineering
Industries Federation of Southern Africa (SEIFSA), which is 76 years old this year, is
the undisputed voice of the M&E cluster of industries.

Given SEIFSA's important role as a champion of manufacturing, I hereby nominate
SEIFSA Chief Executive Officer Kaizer M. Nyatsumba for membership of the BRICS
Manufacturing Work Stream. Attached hereto is a copy of Mr Nyatsumba's Curriculum
Vitae.

We would be most grateful if Mr Nyatsumba's nomination would receive your favourable
consideration.

Yours Sincerely

Elias Monage
President and Board Chairman

About three months or so passed before he returned to me to say
that the BBC had agreed to his recommendation that I, as SEIFSA
CEO, should chair the structure's MWG because our federation
was involved in manufacturing. He told me that BBC Chairperson
Busi Mabuza would send me a letter of appointment, whereafter
we would be able to set in motion the handing over of the
chairmanship from Nizam Kalla, owner and Managing Director of
Amka Products (Pty) Ltd, to me. Kalla would then become Deputy
Chairman of the MWG.

While we were still awaiting that official communication
from Mabuza, Monage asked me to meet him at Tashas
Restaurant at Nelson Mandela Square in Sandton one morning.
During that meeting, which started at 8am, he told me that he was
expecting Mabuza to sign my appointment letter at any time and
to have it sent off to me. However, it was what he said next that
shocked me and revealed a little more of the man: as Chairman
of the BBC MWG, he said, I should feel free to take advantage of

any opportunities that may exist to conclude my own business deals. I was stunned, but I did not respond to that piece of advice. It has never been in my nature to take personal advantage of opportunities to which I was exposed as part of my work, but it was clear that for Monage that was par for the course.

I thought back to two crucial meetings that we had held with key stakeholders during Monage's first year on the Board. One was with the ANC's Economic Transformation Committee (ETC), which was headed by Enoch Godongwana, and the other was with NUMSA General Secretary Irvin Jim during the 2017 wage negotiations. I arranged meetings with the ANC's ETC, as an important economic policy-making arm of the governing party, from time to time to create an opportunity for us to lobby it. I always took the Chief Economist along with me to these meetings, as well as selected Board members. Invariably, these included the Board Chairperson, one of the Vice-Presidents and Board members from KwaZulu-Natal and the Western Cape to ensure that all regions were represented. That was important to ensure that we did not have a big SEIFSA delegation arriving at a meeting with stakeholders.

The Chief Economist and I always worked on the presentation we wanted to make at the meeting with the committee, and I then shared it with those who were selected to be part of the meeting in order to obtain their input. The intention was to ensure that we used such opportunities to raise the concerns of the broader M&E sub-sector and not only those particular to the companies to which members of the delegation belonged. On the occasion in question, the Board had chosen Pimstein, Pieter du Plessis and KwaZulu-Natal Engineering Industries Association Chairman Ian Delport. However, just before the meeting with Godongwana, ANC Secretary-General Gwede Mantashe and their colleagues started, Monage – who had not been selected to be a member of the SEIFSA delegation – arrived out of the blue to be part of our delegation.

The same happened when Pimstein, Trentini, one other member of the SEIFSA Negotiating Team and I had scheduled a meeting with Jim and his colleagues at the NUMSA head office.

Again, without having been chosen to be a member of the SEIFSA delegation, Monage showed up unannounced. It seemed to me that there were stakeholders that he considered quite important, and by whom he wanted to be seen to be part of the SEIFSA leadership team. At the time, I found that behaviour strange, but I was not overly bothered by it, and nobody subsequently raised it because at least he did not break rank to champion his own cause.

On 29 July 2019, Monage and I held a meeting with CEOs of selected companies within the SEIFSA fold to introduce the MWG to them. At his suggestion, I had invited 50 CEOs to the online meeting, during which we solicited their input on matters they wanted South Africa's MWG to take up with its counterparts from Brazil, Russia, India, and China. Among the participants in the meeting were then Atlantis Foundries CEO Mervyn Moodley, SAAB President and CEO Trevor Raman, Defy CEO and Regional Director Evren Albas, Columbus Stainless Steel CEO Lucien Matthews, and Denel Aeronautics Senior Marketer Lesetja Mogoba. On 4 September 2019, SEIFSA Chief Economist Dr Michael Ade, Economist Marique Kruger and I attended a MWG hand-over meeting at Kalla's office in Centurion. Also present at the meeting were BBC members Elias Monage and Dr Stavros Nicolau, with the latter being the person within the BBC responsible for the manufacturing sector.

The hand-over meeting took place over lunch, with Kalla a gracious and generous host. Everybody attending the meeting left with a gift; as usual, I did not keep mine. Instead, I declared it and surrendered it to SEIFSA for its use, as is consistent with the federation's policy. Ade was going to assist me in driving the work of the BBC MWG, while Kruger was going to be a vital part of the team by providing secretarial services. From then onwards, hosting of monthly MWG meetings was going to be SEIFSA's responsibility – and I knew that we could not beat Amka's generosity.

We got our first chance to host the MWG meeting in the SEIFSA Boardroom on 10 October 2018, during which we had a full briefing on the working group's work and priorities until then. We were also made aware of the fact that we were working towards the 2019 summit of the BBC, which would coincide with the BRICS summit to be hosted by Brazil that year. Nicolau and Monage attended some of our meetings over the next few months, and the BBC approved our proposed plans for the MWG summit in Brazil.

At his first Board meeting in November 2018, following approval of the minutes of the previous Board meeting, Tsehlo asked for details regarding Michael Pimstein's resignation from the previous Board. When neither Monage, as Chairman, nor Bridgette Mokoetle, as Company Secretary, responded to his question, I explained that Pimstein had stepped down because of differences he had with the rest of the Board. I thought that such a broad explanation was adequate because it related to a matter which predated Tsehlo's appointment, but he pushed for more details. When none was forthcoming from me or anyone else, he stated – on the record – at the end of the meeting that he was not happy with the response that had been given and alleged that I had withheld important information. His reservation was noted in the minutes.

At his second Board meeting on 1 February 2019, during an annual strategy session, Tsehlo asked if SEIFSA was "primarily a products and services organisation or an employer's organisation leveraging its voice". We took him through the strategic options previously offered to the SEIFSA Council and explained that the associations had come out in favour of the "hybrid option", which required the federation to raise revenue through the provision of products and services in addition to being a collective bargaining agency and to doing advocacy and lobbying work. Although he understood the strategic choice made by the SEIFSA Council, Tsehlo said he disagreed that the approach could legitimately be referred to as "a hybrid option". Again, that was his view, and none of us felt any need to respond to it.

Early in January 2020, while I was still on leave, I received a call from him inviting me to a meeting with Dr Bernard (Bernie) Fanaroff at the Department of Trade, Industry and Competition in Pretoria to discuss the Steel Master Plan (SMP). SEIFSA Chief Economist, Dr Michael Ade, and I had already attended the preliminary meetings on the SMP with Minister Ebrahim Patel and other stakeholders in Pretoria, and we had already invited submissions from SEIFSA-affiliated employer associations and their member companies and had woven them into our comprehensive submission to the Department. Patel had since appointed Fanaroff as the SMP Coordinator, and he had contacted his past acquaintance, Monage, to schedule an early meeting with SEIFSA. I dragged myself to that meeting.

Over the next few months, Fanaroff convened several meetings, almost always online, to help him to understand and better appreciate the M&E sub-sector so that he could draft the SMP. There were meetings scheduled with the SEIFSA constituency for which Monage randomly chose people he wanted involved. From the board, Tumi Tsehlo was always one of them. Over time, Monage also included Andrea Moz and Nonhlanhla Ngwenya. Other consultative meetings involved stakeholders from the broader M&E sub-sector, with SEIFSA – which represents all parts of the diverse industry – again being one of the participants. Officially representing the SEIFSA position were Dr Ade, our Chief Economist, and myself.

Although Minister Patel had promised swift progress towards the completion of the SMP, in reality the process proceeded at a snail's pace, with much to-ing and fro-ing. To some extent, that was not surprising, given the very diverse nature of the M&E sub-sector. As invariably happened, a proposal that addressed the concerns of one part of the sub-sector often caused concerns to other sub-industries or another sub-industry, and that resulted in the absence of consensus in the broader SMP meetings convened by Dr Fanaroff. Consequently, it took more than a year for the first draft of the Plan to be completed.

As the process dragged on, frustration grew among the various stakeholders and rank opportunism reared its ugly head in relation to others, among them certain Board Members. These men began to use some of the meetings with Fanaroff, convened ostensibly to discuss SEIFSA's position on the SMP, to promote their respective companies and to campaign for both business opportunities and BEE partnerships. In one such meeting, a Board

Member complained on record that, as a black-owned company, KLM would not benefit appropriately from the SMP as it stood.

I was so deeply concerned about this Board Member's blatant abuse of a SEIFSA platform to promote the interests of his company that, after the meeting, I raised my concerns privately with Monage, who was SEIFSA Vice-President at the time and often attended the Microsoft Teams meetings from Trentini's old office. Alpheus Ngapo was SEIFSA President at the time. Despite his more senior position, Ngapo was not part of the SMP meetings with Fanaroff. He continued to hold the view that, based on the input we were receiving from member companies, Ade and I were best placed to represent the federation's interests. That was not the case with certain Board Members who, as BEE players, had a vested interest in the outcome of the SMP process.

Monage would also act outside the authority of the Board of Directors. Following the Board's approval of SEIFSA's budget for the 2020/21 financial year, which included a 6% adjustment in membership fees, the new Executive Director of the CEA, Anthony Boy, wrote to me to ask for a discount for his association, which was by far the wealthiest among the affiliated associations. It seemed to me that he was intent to demonstrate to the CEA that he could take SEIFSA on and get a special dispensation for the association. In a meeting with Boy and some members of the CEA Executive Committee, I explained that the budget had been carefully considered by the Board of Directors and that I did not have the authority to reduce membership fees for the CEA. When we could not resolve the issue, I promised to escalate the matter to the Board for its consideration.

Indeed, I subsequently wrote to the Board to apprise it of the CEA's position and asked that an urgent meeting be convened to discuss the matter. However, I heard not from the Board, but from Monage, who informed me that he would meet with the CEA to discuss the issue. He subsequently informed me that he had agreed to a 4% membership levy increase for the CEA, instead of the 6% decided upon by the Board for all associations and their members. When, at the next Board meeting on 3 August 2020, Moz asked why a Special Board Meeting had not been convened to discuss the CEA membership fee matter, Monage replied that "the matter was resolved". He explained that the CEA's membership levy for the 2020/21 financial year had, indeed, been adjusted downwards, "taking into account circumstances in the construction industry", but he apologised for having failed to inform the Board accordingly.

I was very concerned about the way Monage had dealt with the issue as it had the potential to set a precedent for the CEA and, indeed, other member associations.

When a draft SMP was finally produced towards the end of 2020, I circulated it to affiliated employer associations and member companies and asked them to send their comments to Ade within a certain period. Once we had gone through the internal feedback, we compiled comprehensive feedback on behalf of the federation and copied the Board in on the submission we made to the Department of Trade, Industry and Competition.

Monage and I were to differ sharply on our approach to the SMP. As a matter of principle, I was guided by the clarification we had made to the federation's MoI back in 2014; this states that the SEIFSA Council "shall be responsible for those affairs of the federation that require a mandate from member associations". My firm view, therefore, was that, as CEO, I was obliged to solicit input from member associations and to submit to the Department of Trade, Industry and Competition only what amounts to the mandated position of the SEIFSA Council. Supported by Tsehlo, Monage, on the other hand, insisted that the Board should have the final word on SEIFSA's position on the SMP.

Following differences on this matter in the November 2020 Board meeting, I asked Acting Company Secretary Louwresse Specht to scour the MoI and to produce a legal opinion on it. In her report in the Board meeting of 1 February 2021, Specht confirmed my understanding of the MoI, concluding that the SEIFSA Council "has the authority to collaborate with the federation in making submissions on the Steel Master Plan", while the Board "is merely entitled to enquire on the status of submissions on the SMP, but Board Members are not automatically entitled, in their capacities as Board Members, to have their submissions on the SMP considered".

Upon receipt of the copy of our comprehensive response to the Department of Trade, Industry and Competition, Tsehlo wrote to me on 23 October 2020, copying in Ngapo and Monage, and sent me his own suggested additions to our response. "I had understood," he wrote in his email, "that the Board would try to discuss this before the response is submitted to the DTIC." He said that although he had gone through our submission and concurred with it, nevertheless he believed that the submission would benefit from the "additional points" that he wanted to make.

In my reply to Tsehlo on the same day, I wrote:

> Please note that this is a matter which required input from affiliated associations and their member companies, and not from the Board. The SEIFSA Council deals with matters of mandate, on matters where we have to represent the views of the federation as a whole, while the Board deals with matters of governance.

> However, there was nothing which prevented members of the Board, in their individual capacities (and not as Board members), from sending us their own inputs for consideration for inclusion in the official SEIFSA submission, provided they appreciated the fact that membership of the Board did not give them a bigger say in such matters than the other member companies. It is very important that we are always mindful of the responsibilities of the two structures (i.e. the SEIFSA Council and the Board), and that neither encroaches on the other's territory.

> We will be sending our official submission on Monday, 26 October, which is the deadline for submissions. That means that, although we had asked that all inputs should reach us not later than 12 noon on Wednesday, 21 October, we still have room to consider yours, on behalf of your company, and not as a Board member, between now and then.

On 26 October 2020, Tsehlo wrote back to thank me for my response. He said that he agreed with my "assessments and points regarding the contribution," stating that it was "the correct manner of approaching it".

In the Board Meeting of 1 February 2021, I reported that I had made contact with Fanaroff to make enquiries about the progress of the SMP following our comprehensive submission after the release of the first draft. Fanaroff had informed me that Minister Patel was not happy with some of the submissions from the industry, but that he would soon be publishing the Plan in the *Government Gazette.* Unbeknown to us at the time, Monage had been continuing discussions with Fanaroff on the SMP. Following my report to the Board, Monage told us that Minister Patel had appointed 42 individuals from various companies, including himself, to be on the SMP Council. He revealed that the first meeting of the SMP Council was scheduled to take place on 8 or 9 February 2021. He said that he had not received the identities of the other members of the SMP Council, but that he had requested a list.

This was shocking to me. It was clear that Monage's direct involvement in the process, because of his personal relationship with Fanaroff, had worked in his favour as an individual, and not in favour of the organisation on whose Board he served. I informed the Board that SEIFSA, which Minister Patel had earlier earmarked to play a secretariat role for the SMP, with the Department and the industry paying for such services, had not received calls for nominations for the SMP Council. When I asked if other Board members knew who else was appointed onto the SMP Council, three people (among them Moz) – the same people that Monage had personally involved in the SMP meetings, without any reference to the Board – disclosed that they had also been appointed to the SMP Council. Clearly, Monage had succeeded spectacularly in building a faction beholden to him within the Board!

Ngapo, as Chairman, expressed concern that SEIFSA, as the voice of the M&E sub-sector, had no one from its ranks appointed to the SMP Council. He felt that SEIFSA would be left out of participating on the final version of the SMP, with only the voices of selected industry leaders – pushing their own agendas – being heard. No response was given when Ngapo asked the SEIFSA Board members who were appointed to the SMP Council if they would use their new positions to advance SEIFSA's agenda or their own companies' agendas.

Certain Board Members had begun to use the online meetings with the Department of Trade, Industry and Competition to promote their own business interests. Increasingly, the two men used their membership of the SEIFSA Board of Directors – and the access that it gave them to policy-makers – to promote their selfish business interests rather than to advance the interests of the federation and its stakeholders. When my private efforts to get them to desist from such conduct failed, I reported the matter to Ngapo, in his capacity as Board Chairman, and he felt that the matter should be placed on the agenda of the May 2021 Board Meeting.

Interestingly, more than two years later, Monage (2023) – who had been such an enthusiastic proponent of the SMP – was to complain bitterly in the media that the SMP's "grand aspirations to re-industrialise the steel industry are beginning to disintegrate under our own watch". He complained about Patel's alleged inaccessibility, claimed that the SMP's lofty goals were "becoming increasingly illusive" and averred that Patel and team were implementing policy "in a fragmented manner, with a short-term view and with pockets of industry being pitted against one another". Monage (2023) declared that the industry was "quickly losing faith in the SMP process".

"Industry simply cannot continue to invest the amount of time, effort and resources as it has done thus far for altruistic reasons, particularly when the experienced reality is one of a continued deterioration of the business and operating environment, company closures and job losses," Monage (2023) wrote.

Another point of difference with Monage, which led to considerable estrangement between us, was his gratuitous advice to me, towards the end of his tenure as President, around September 2020, that I should appoint Nuraan Alli and Louwresse Specht as Sales, Marketing and Communications Executive and IR and Legal Services Executive (and Company Secretary), respectively. I had appointed the two women in those positions in acting capacities some months earlier when vacancies arose – following the resignation of Sales, Marketing and Communications Executive Jackie Molose and IR and that of Legal Services Executive Sibusiso Mthenjana in May 2020 – to give them a chance to show that they could perform at that level. Alli had been the Sales Manager for several years and Specht had joined us in January that year as IR and Legal Services Manager.

I was keen to appoint Alli into Molose's position, but I had explained to her that, before I could consider appointing her in the role on a full-time basis, she needed to demonstrate that she could constructively lead a diverse team and that she could generate sponsorship for the Southern African Metals and Engineering Indaba and the SEIFSA Awards for Excellence. During her time with SEIFSA, Alli had had fractious relations with both her direct reports and her peers. Various interventions were made, including external management coaching, to help her improve her people management skills, and she had subsequently shown the necessary improvement. However, over the years she had failed to raise sponsorships, which was a very important part of the job to which she aspired.

Specht was barely three months into her position as IR and Legal Services Manager when Mthenjana, who had employed her, resigned to join a member company. Given her relative youth and the fact that she was still new at SEIFSA, I appointed her to the higher position in an acting capacity, without making any promises to her about her prospects for the job. I was planning to watch her performance over a six-month period and then to make a decision about her appropriateness for the position.

I was surprised, therefore, when, out of the blue, as we were making plans for the 2020 AGM, Monage came to see me and advised me to appoint the two women into those positions on a permanent basis. As I had done when he advised me to use my position as Chairman of the BRICS MWG to conclude personal business deals, I merely listened to his advice, but did not respond to it. During my seven years with SEIFSA, that was the very first time that any Board member had sought to involve him- or herself in the appointment or promotion of staff members.

Specht performed well as Acting IR and Legal Services Executive and Company Secretary, so I appointed her to the position on a permanent basis with effect from 1 January 2021. With Alli still not having raised any sponsorship for our corporate events, I extended her period as Acting Sales, Marketing and Communications Executive for another six months to give her another chance to prove herself.

The final straw that broke the camel's back in my relationship with Monage must have been the complaint that was laid with me by a member of our EC Division. I was in my office in March 2021 when she walked in, almost in tears, to complain that she had just got off the phone with Monage, who had been rude to her. She told me that Trentini had given her Monage's number and asked her to call him because he wanted to talk to her. When she called him, Monage had allegedly demanded a free copy of *SEIFSA's Price and Index Pages*, a publication which was sold via subscription, on the pretext that he was "working on a SEIFSA strategy". She informed me that when she told Monage that she could not send him a copy of the publication because he was not a subscriber, he was verbally rude to her.

In my CEO's Report for the Board Meeting of 3 May 2021, I wrote:

> The allegations, in the words of the staff member in the EC Division, are contained in an email which she sent to me, and which was sent to the Company Secretary for inclusion in this Board Pack. I do believe that this is the right forum at which this matter should be discussed so that we have complete understanding of what transpired during that telephone call. Of particular concern to me is the fact that there is a standing Board protocol that Board members will deal only with the CEO on all matters, as a result of which the Board has previously sent out communication to staff

members indicating that none of them should ever deal directly with Board members on any issues.

Therefore, item 10.1 on the agenda of the Board Meeting of 3 May 2021 was called "Complaint from EC Division Staff Member".

Monage must have been very unhappy upon receipt of that Board pack. Not only were the tendencies to abuse SMP meetings with the Department of Trade, Industry and Competition for the benefit of their respective companies going to be discussed at the meeting, but so, too, was a complaint laid against him by our Economist. Clearly, he had to act pre-emptively to prevent that from happening.

## Guiding Principles for SEIFSA's Board of Directors

Although corporate governance was observed in the breach at SEIFSA, with the organisation traditionally having had an "Executive Committee" instead of a Board of Directors, much progress was made from February 2014 to institutionalise good corporate governance. Not only was the MoI revised to ensure that the SEIFSA Council and the Board of Directors had clear and mutually reinforcing roles, but a Board Charter was also adopted and published prominently on the federation's website, and a Delegation of Authority Matrix was approved.

Not only do the revised MoI and the Board Charter clearly delineate the roles of the CEO and the Board Chairperson, but they also make it clear that the CEO is the primary spokesperson for the organisation, with the Chairperson being the primary spokesperson for the Board. Henk Duys, who was President when I joined the federation in 2013, had made the same point in his Chairman's Report in the 2013 Annual Report: "The President's job was simple: he chaired the meetings and occasionally made public comment, but the real spokesman and leader was the Executive Director who operated within clear mandates and policy frameworks set for him by the SEIFSA Council and [the] Executive Committee", Duys had written. The federation had also since compiled General Principles which are sent to all Board nominees and the member associations. For the sake of completeness, the General Principles are presented below:

1. The Steel and Engineering Industries Federation of Southern Africa (SEIFSA) is an industry body which exists to serve the interests of the metals and engineering sector in Southern Africa in its entirety, and not some section of it.
2. SEIFSA exists to serve the interests of all its members, large and small, and not only of those with the loudest voice or the biggest size.
3. All member associations have an equal voice at SEIFSA meetings, regardless of their size.
4. While every attempt will be made to reach consensus on matters that require a mandate and fall within the purview of the SEIFSA Council, where differences exist among member associations, opposing views will be noted and recorded but the majority decision taken by member associations will be SEIFSA's position and be binding on member associations.
5. In keeping with Good Corporate Governance, the CEO and his Executive Team are responsible for SEIFSA's operations and are accountable to the Board as constituted at the federation's Annual General Meetings.
6. The SEIFSA Board is responsible for the oversight of the federation, including the approval of its strategy and budget, and is accountable to member associations through the Annual General Meeting.
7. In keeping with Good Corporate Governance, once elected, all SEIFSA Board Members will have a fiduciary duty to the federation, respect the Board Charter and serve only its interests, and not those of their respective companies or associations.
8. The SEIFSA Board will endeavour to be representative in terms of race and gender, with at least three of the seven elected Non-Executive Directors (NEDs) being black (African, coloured or Indian) and at least two of the seven NEDs being women.
9. All elected representatives of associations affiliated to SEIFSA are required to advance the interests of their associations and SEIFSA at all times and never to consort or cooperate with

or leak information to bodies and/or organisations opposed to the federation. Representatives of affiliated associations found to be in violation of this provision will be excluded from SEIFSA meetings and their associations will be requested to send other representatives to the federation's meetings.

10. All members of SEIFSA who are conflicted on any subject up for discussion at Board or Council Meetings will be required to recuse themselves during the course of that discussion and rejoin the meeting once that discussion is over.

## Conclusion

In this chapter, the degree of lack of alignment between some members of the Board and myself has been shown to have become more pronounced. At the heart of the non-alignment was pursuit of self-interest, at a time when all Board members should have been faithful to their fiduciary duty to SEIFSA and all its stakeholders. From the content of this chapter, it is clear that some individuals considered their membership of the SEIFSA Board of Directors to be an opportunity to be leveraged to advance their own selfish interests, and that my opposition to this inexorably led to our parting of ways. To accomplish that goal, the self-same individuals had to resort to questionable methods that ended up costing SEIFSA dearly at a time when it was still trying to come to terms with the devastating financial effects of the COVID-19 pandemic.

# Chapter 14

# An Engineered Parting of Ways

## Introduction

Although the degree of non-alignment between me and some key individuals within the SEIFSA Board was more pronounced by the end of 2020, it did not possibly occur to me that their determination to prevail would lead to the type of recklessness that it unleashed. As things turned out, two fortuitous events were to work in their favour.

As will become clear below, internal ambition on the part of some senior – and until then trusted – members of my Executive Committee, combined with the desire of some Board members to use SEIFSA as a vehicle to advance their business interests, proved to be an extremely potent combination.

## Internal Ambitions

During my interview in September 2013, then SEIFSA President Henk Duys let slip that the job had been advertised internally within SEIFSA and among the federation's member associations, but that there had been no suitable candidate available. He told me that one applicant had come from within SEIFSA, but the panel had felt that that individual was not yet ready for the position, hence he was not invited for an interview.

Over the next few years, I was to come to the conclusion that the person mentioned by Duys during my interview had been Operations Director Lucio Trentini. Within SEIFSA, the route to the top position appeared to have been through the Operations Director position: my predecessor, David Carson, had been Operations Director for many years. It made sense, therefore, that Trentini – who had joined the federation two years after graduating with a Bachelor of Arts degree from the University of the Witwatersrand – would aspire to see his career take the same trajectory.

However, like Duys and team, I also concluded that Trentini was not ready for appointment to the helm of the organisation at the time. He was more an IR expert than an all-round Manager and often defaulted to IR work, much to the disadvantage of the then IR Executive, his protégé Gordon Angus. The rest of the business did not seem to interest him. It took a lot of work on my part to drag him away from IR work and to involve him in strategic work. That did not come naturally to him, but over time he grew into the role.

When we collectively engaged in succession planning in my first two years in the job, we could not identify a possible internal successor to me. Over time, Trentini emerged as a possible successor, but he still needed time to be fully ready to walk into the job. His willingness to learn and to work closely with me as a strategic support meant that over time he reached the "ready now" stage for succession. But so, too, had CFO Rajendra Rajcoomar, who had the advantage of having a Master of Business Administration from the Wits Business School and of having worked in senior positions at other companies. He had the advantage of understanding finance and accounting, and he was a more strategic business operator.

In subsequent annual succession plans, Trentini and Rajcoomar placed first and second, respectively, as potential successors in the CEO role, with the gap between them growing increasingly negligible. I worked closely with both men and relied on them as key lieutenants.

At the end of each year, as we got ready to wish each other well during the festive season, Trentini often remarked on my staying power. "Ah, you have finished yet another year in the job, Kaizer," he would say. "You are so resilient; you have weathered storms and defeated all efforts to cut your tenure short." In response, I would tell Trentini that although there would always be people who would like to see one's back, it is only the individual concerned who would give ammunition to those individuals by doing anything that they would be able to seize upon to advance their agendas. In my case, I told him, there was absolutely nothing that they would be able to use to achieve their goals because I was always guided by my principles in everything that I did.

In September 2020, Trentini celebrated his 30th year with SEIFSA. In recognition of his long service, I asked that he be featured prominently on the cover story of our bi-monthly publication, *SEIFSA News*. Although, to save costs, we no longer published physical copies of the magazine, on that occasion I asked for 20 copies of *SEIFSA News* to be published and given to him as a souvenir.

Exactly three months younger than me, Trentini would have been worried that each year I spent in SEIFSA's employ reduced his own chances of succeeding me. At the beginning of 2021, I was 58 years old – and he turned that age on 2 February 2021. With SEIFSA's retirement age being 63, that meant that his chances of being CEO were getting slimmer.

I was fully aware of this dynamic, but I also knew that I had no intention of staying with SEIFSA until my retirement. I was just about to finish and submit my PhD thesis at the time and intended to move on to a consulting role and Non-Executive Directorships upon graduation. However, neither he nor Rajcoomar knew of my plans.

Although Rajcoomar reported to me, he started getting too close to Trentini and ran everything by him before coming to me with any proposal or report. For a while, the two of them convened weekly meetings with Executives and Managers to plan revenue generation. This was Rajcoomar's idea, and I supported it, not reading much into it. A year earlier, I had asked him to open up to tender the running of the SEIFSA Training Centre. When GijimaAST's five-year contract expired in 2015, I had reluctantly renewed it because I had not had enough time for us to call for tenders from interested companies. In 2018, I served notice to GijimaAST that, when the contract expired in 2020, we would put it out to tender to test the market – I had assigned this task to Rajcoomar as CFO. He, in turn, put together a team which included Trentini and Sales Manager Nuraan Alli, among others, and they recommended a company called Thuthukisa for appointment, on financial terms more favourable to SEIFSA.

For some time, we had a policy that limited the value of gifts to be accepted by staff members to R500 each year, with all gifts to be declared in the Gifts Register. The Human Capital and Skills Development Executive was the custodian of the Gifts Register and presented a quarterly report to the Executive Committee on the gifts received, kept and/or donated to SEIFSA. As a matter of habit, I routinely declared all gifts and then donated them to SEIFSA, very rarely keeping anything for my own use. Others were in the habit of keeping gifts once they had declared them.

As always, at our Executive Committee Meeting on 22 January 2021, Human Capital and Skills Development Executive tabled her Gifts Register for the October-December 2020 period and for the 2020 calendar year. The latter revealed that, during 2020, Trentini had accepted gifts worth more than R500, which was against our policy. He then transferred to the Human Capital and Skills Development Executive the responsibility for compliance with the policy, asking her to keep a running tally of the value of the gifts that he had received to avoid the kind of situation that he then found himself in. The situation called for disciplinary action, but I asked Trentini and other Executive Committee members to ensure that not only was there no repeat of what had occurred, but that their respective team members complied with the policy fully.

In retrospect, it would appear that there was more to the closeness between Rajcoomar and Trentini than met the eye. This became clear when we were working on the 2021/22 budget, to be presented to the Board on 3 May 2021. As usual, we had held a series of budget discussions internally and had discussed areas where cuts could be made to ensure that we recouped the loss suffered in the 2019/20 financial year due to the COVID-19 pandemic. Before we settled on yet another limited round of retrenchments, Rajcoomar and Trentini had proposed that we give up the west wing of the 6th floor we occupied and squash everybody into the east wing. I considered such a proposal irresponsible: we could not reduce physical spaces among staff members and have some of them even sharing offices at a time when the country was still battling COVID-19. Therefore, I rejected that proposal, and we resolved to go with the second round of retrenchments. Our calculations showed that we would be able to present a break-even budget.

Early in March 2021, I was diagnosed with a terrible bout of shingles. Not only did I have to be hospitalised, but I also had to bear agonising pain at home, unable even to wear a shirt. That period of absence from work was the first event that offered the conspirators marvellous time to plan at will. As my key confidant, Trentini was fully aware of my condition. After my initial recuperation, I had to work from home because I could not wear anything on my back. I was surprised, when we had an online meeting to finalise the 2021/22 budget, when Rajcoomar again placed on the table the suggestion of us foregoing the west wing of the 6th floor. As far as I knew, I had made my opposition to that plan very clear, which meant that it should not again have been listed as an option available to us.

That was very strange because Rajcoomar was a man who was normally attentive to details. Never before had he pushed against a decision that I had made once I had articulated it after we had discussed an issue. That showed that something was amiss. I reminded him that I had made my view on that proposal clear during our budget discussion in the office before I fell ill in March, and I insisted that he factors into the budget the retrenchments already agreed upon.

That year, for the very first time, Moz insisted that the budget be presented to the Audit and Risk Committee on the morning of the Board meeting before it was to be submitted for the Board's consideration later that day. That meeting, which I joined online, was unlike any other Audit and Risk Committee meeting that I had attended before. A Chartered Accountant who came alive when finances were discussed, Moz was incredibly petty that day. He questioned not only all the budget assumptions, the revenue assumptions and the planned costs, but also Rajcoomar's financial accounting competence. Although he had twice before approved SEIFSA budgets as a Board member, suddenly it appeared as if it was the first time that he had come across the federation's budget, which was prepared on the same template.

Interestingly, Moz asked me why I had not favourably considered the proposal for SEIFSA to give up the west wing of the sixth floor as a cost-saving opportunity. In response, I explained that, as an Executive Committee, we have always had a surfeit of cost-containment measures to consider, and that we always took the option that we considered the most appropriate or the least harmful at the time. I explained that I had turned that proposal down because COVID-19 was still a reality and that I could not, with a clear conscience, approve a proposal which would have the effect of endangering staff members' lives.

At 11:25, as he concluded the meeting, a very adversarial Moz stated that he would advise the Board not to approve the proposed break-even budget, which was based on such cost-containment measures like retrenchments.

The Board meeting was due to start at 11:30. Little did I know that a bigger surprise awaited me at that meeting.

## Certain Board Members Strike Back

While I was still recuperating at home in April, Board Chairman Alpheus Ngapo called to tell me that he had decided to resign from the Board owing to other personal commitments. He said that he no longer found it easy to travel from Pretoria to Johannesburg for Board meetings. The news of his resignation hit me like a bolt from the blue. Alongside Neil Penson, he was by far the steadiest pair of hands and the most level-headed person on that Board, and I knew that his departure would not be good for the federation. Filled with a sense of foreboding, I asked Ngapo to consider staying on at least until the 3 May 2021 Board meeting, but he insisted that he could not devote an extra day to SEIFSA matters.

To this day, I have no idea if there was more to Ngapo's resignation. Like a man on a mission, Elias Monage – for whom

Ngapo had stood aside a year earlier – moved swiftly to convene a Special Board Meeting to lobby to be elected Interim President.

At 11:30, when the Board meeting was due to start, I logged in on Zoom and found that the Company Secretary, Louwresse Specht, was not there to let me in. After trying again some minutes later, I called Specht to establish what was happening. When she did not answer her phone, I called Trentini. He, too, did not answer at first. After a while, he called me back to tell me that there was a meeting of Non-Executive Directors (NEDs), that had been convened by Monage as Interim Chairman.

I kept checking with Trentini and Specht whether the meeting of NEDs was over, and they undertook to call me when it ended. It was not until 12:30 that I received a call from Specht, who told me that the meeting had finished and the Board meeting was about to start, so I should join it. I duly did so.

At the beginning of the meeting, Monage welcomed everybody and asked if there were proposed amendments to the agenda. Tsehlo said he would like to add to the agenda a matter he referred to as "the CEO's Conduct." I was taken by complete surprise and did not know what he referred to. It was then that I realised that the previous meeting had most likely been about me. It was clear that it had been an opportunity for Monage and Tsehlo to convince the other NEDs to support their proposed course of action against me. Asked where he wanted that item added on the agenda, Tsehlo said he would like it right at the beginning of the meeting, after the safety moment. (One of the things I introduced in 2014 at SEIFSA was a safety moment at the beginning of all meetings, which was an opportunity to reflect of matters of safety). That meant that the new agenda item would be before the consideration of the Minutes of the Board Meeting of 1 February 2021 and before the CEO's Report. His proposed amendment to the agenda was accepted and the agenda was adopted.

When we got to that agenda item, Tsehlo said the following things:

- He had "lost confidence in the CEO's leadership;"
- "The CEO was defensive and was showing distractive behaviour;"

- He "suspects that there was bullying of staff members;"
- The Board's inputs "were ignored;"
- On the issue of SEIFSA's mandates and relevance to its members, "the CEO has refused to follow the Board's direction;"
- He was informed that Alpheus Ngapo had resigned as SEIFSA President and Board Chairman "because he could not work with the CEO;"
- "There was no action plan from the strategy session (held a month earlier) and there was no plan to respond to the Board's efforts;" and
- "SEIFSA's financial performance was weak over the past two years, but there has been nothing done by the CEO".

Monage added that "a number of people had resigned from the Board, and the information was not communicated to the Board". He said there was a need to suspend the CEO. Tsehlo supported the motion, saying it was important that "a precautionary suspension of the CEO takes place to allow the Board to investigate the state of the organisation because I do not believe the picture that is communicated to the Board is accurate".

The motion to suspend the CEO was carried, and Monage said a formal letter of suspension would be sent to me that afternoon. I was then excused from the meeting, and I duly exited it.

That afternoon, Trentini called me to say that he was aware that "there was a plan to oust you" (but he had not told me about it), that my suspension had been communicated to the SEIFSA Council, whose meeting started at 14:00, that he and CFO Rajendra Rajcoomar had jointly been placed in charge of SEIFSA "because they can trust us to do what they want", and that he was about to have a Staff Meeting to inform members of staff about my suspension. During the meeting with staff members, they were all informed not to contact me and, in the letter that I subsequently received that afternoon from Monage (through Specht), I was also ordered not to communicate with any SEIFSA staff members.

On 11 May, I received an email from one Paul Boughey, CEO of Resolve Communications. Although sent to my personal email address (which was itself strange), the email was written to me in my capacity as SEIFSA CEO. It contained a proposal for SEIFSA to work with several business associations on a "Save Our Ports" campaign. I did not reply to that email immediately. Upon thinking about it, I realised that this was most likely a trap for me; I thought the Board or the individuals currently running the federation had spoken to Paul and wanted to see if I would reply as SEIFSA CEO. The following day, I replied to him as follows:

Dear Mr Boughey

Thank you for your e-mail.

I am not at work at the moment. Please direct your e-mail to Operations Director Lucio Trentini (Lucio@seifsa.co.za), who is currently standing in for me.

For a full month after my suspension, I did not hear from SEIFSA, and no charges were preferred against me. Instead, an external investigator was appointed to interview those members of staff who reported to me, with the women allegedly asked, among other things, if they had ever felt harassed by me by virtue of their gender.

About two weeks after my suspension, Ngapo called to say that he had just heard about my suspension and that he was sorry about it. He said had he known that certain Board Members were cooking this plan, he would not have resigned as President and Board Chairman. When I told him that it was alleged during the Board Meeting that he had said he had resigned because he could not work with me, he said that was a lie. He said he and I had come a long way (he was on the SEIFSA Board when I joined the organisation in 2013) and that he had worked well with me. He assured me that if he had been unhappy about anything, he would have been able to raise it directly with me.

Not long after I was paid out to part ways with SEIFSA, Trentini was appointed CEO and Rajcoomar left the organisation. Trentini subsequently lured former Senior Economist Tafadzwa Chibanguza away from the Minerals Council and appointed him Chief Operations Officer.

## What are the Facts?

SEIFSA's MoI gives certain powers to the SEIFSA Council and others to the Board of Directors. Everything that requires a mandate or that would see SEIFSA pronouncing publicly on a particular issue of importance on behalf of the associations requires a mandate from the SEIFSA Council, while the Board's responsibility is as stated in the Companies Act 71 of 2008. The Board is responsible for strategy, policies and general oversight of the organisation, through the CEO to whom certain responsibilities have been delegated through the Delegation of Authority and the Board Charter. However, our BEE entrepreneurs had wanted the Board to usurp the responsibilities of the SEIFSA Council, and I had opposed that as a matter of principle.

Secondly, as has already been indicated in the previous chapter, certain Board Members had used their membership of the SEIFSA Board to advance their business interests, instead of seeking to advance the best interests of SEIFSA. I had been against what they had been doing. Thirdly, they had this weird idea that Executive Directors are not themselves Board Members who also have a fiduciary duty to SEIFSA. Therefore, whenever I challenged their ideas during Board Meetings, they described that as being defensive.

Fourthly, the allegation that there are some staff members that I have bullied is simply untrue. That allegation would explain why their investigator ran around asking suggestive questions of some staff members. Instead, I have always held staff members accountable and have been approachable and very fair to all of them.

Fifthly, there is not a single Board resolution that I have ignored, nor a single idea resulting from any Board Strategy Session that I have ignored. All Board resolutions have been implemented, and no single Board Strategy Session has ever come up with a concrete idea. By definition, a strategy session is not a Board Meeting; instead, it is an occasion when different options are considered and ideas are debated. Any concrete idea resulting from a strategy session is then presented to the Board of Directors for approval. Therefore, Tsehlo – who had been hostile to me from the time he joined the Board three years earlier – was yet again being economical with the truth.

Contrary to Tsehlo's allegation that Alpheus Ngapo had resigned "because he could not work with the CEO", as I have indicated above, in his telephone call to me Ngapo strenuously denied the allegation, stating that was untrue. He also pledged to appear before the disciplinary hearing as my character witness.

The federation had made growing losses since 2011 when the CBL, which had been a boon to the organisation, expired and was not renewed by the Minister of Labour. When I joined SEIFSA in November 2013, the Board had approved a deficit budget in May that year. I managed to reduce the deficits and eventually turned the organisation around to profitable status in 2017, 2018 and 2019. The losses incurred in the 2019/2020 and 2020/2021 financial years are directly attributable to the COVID-19 pandemic. Taking action against me because of losses suffered during the COVID-19 shutdown periods is hypocrisy of the highest order. These men's own companies registered losses during that period.

Contrary to Monage's allegation that "a number of people had resigned from the Board", as far as I knew only one, Ntobeko Panya, had done so. He had informed Specht that he found it difficult to travel from Cape Town to Johannesburg for Board meetings. When Specht told me about it and asked if she should inform the Board, I told her that, in my seven years at SEIFSA, we had reported on such resignations at the next Board meeting, and did not send out communications as and when such a resignation occurred. Therefore, I advised her to report on that resignation at the May Board meeting. Monage, who knew that was how we had previously done things, opportunistically seized on Panya's resignation to allege that I – and not Specht as Company Secretary – had failed to inform the Board about resignations.

A few days after my suspension was announced within SEIFSA structures, I received an email from Ross Williams, Chairman of the South African Engineers and Founders Association, gleefully celebrating my suspension. The man and I had never before communicated via my personal email address, and yet there he was, using it to write to me. There was no doubt in my mind that he had obtained it from his sources within SEIFSA because I never once took the federation's laptop out of the office. Instead, I left it in the office every day and conducted any business work on my personal computer at home, sending emails to colleagues from my personal address.

## A Hotchpotch of Fabricated Charges

A full month after my hasty, precautionary suspension, I had still not received charges from the SEIFSA Board of Directors. I had to ask my lawyers from Edward Nathan Sonnenbergs (ENS) Inc. to demand that SEIFSA charges me or allows me to return to work. Only then, on 15 June 2021, did we eventually receive the charge sheet, with a notice to attend a disciplinary hearing on 9 July.

Only two of the allegations made by Tsehlo in the Board meeting of 3 May 2021 when I was suspended were included in the charge sheet, which contained a hotchpotch of fabrications. It was alleged that, "during or about 2019 to 2021", I had:

· Used "SEIFSA's e-mail, computer and network resources" to work on my PhD thesis and the subsequent book that was based on it, entitled *Successfully Implementing Turnaround Strategies in State-Owned Companies: SAA, Kenya Airways and Ethiopian Airlines as Case Studies*;
· "Caused and/or utilised programmes used by SEIFSA, including e-mail, laptop and network resources" to work on the aforementioned PhD thesis;
· Instructed SEIFSA employees, "during working hours, to conduct other work in respect of [my] personal and/or business interests" and to work on my doctoral thesis and the book that I subsequently published;
· Used SEIFSA's money to fund purchases for voice-recognition software to further my personal and/or business interests;
· Refused to implement "appropriate corrective action/s to minimize the loss-making situation of SEIFSA, aggravated by the COVID-19 pandemic" by failing to reduce the office rental by forfeiting about 45% of the office space, and failing to consider other cost-cutting measures other than retrenchments;

- Instructed the Company Secretary to register trademarks for SEIFSA, *PIPS* and the SEIFSA Training Centre "in neighbouring countries when the registration of business entities in these countries have not been completed";
- Used SEIFSA's resources "to further the interests of the South African BRICS Manufacturing Chapter, while SEIFSA did not have adequate financial resources to sustain this activity";
- Failed "to comply with SEIFSA's Values and Leadership Declaration"; and
- Become incompatible with the Board by negligently withholding information from it and/or refusing to implement its instructions.

As I read through the charge sheet, one thing became clear: in addition to my objection to Monage and Tsehlo's repeated abuse of their membership of the SEIFSA Board, my completion of my PhD within a four-year period (during which Monage had repeatedly been said to be working on a Master's degree) had struck a nerve. Almost half the charges were related to it!

Yet another thing became obvious: Monage, Tsehlo and company must have known that they had no case against me, but were intent on getting rid of me so that they would be able to do as they wished at SEIFSA. They must have hoped, I thought, that I would not have the stomach for a fight, especially since I was still convalescing from post-herpetic neuralgia, and that I would simply walk away, instead. How wrong they were! Nothing less than vindication of my reputation as a man of integrity, which I had worked hard to build over the years, would suffice for me.

I turn now to the hotchpotch of allegations that were put to me in desperation after six weeks of investigations by an external service provider.

Firstly, it is a lie that I had used SEIFSA's resources or infrastructure to conduct my PhD research. Not only did I never take the SEIFSA laptop out of the building during my entire tenure at the organisation, but I used my own computer at home (on which this book is written) and worked on my studies until midnight each day. To conduct interviews, I took leave (a full day or half a day, depending on the number of interviews scheduled).

The only assistance that I received from my Executive Personal Assistant, Lerato Lebeko (usually during her lunch time) was in the form of establishing interviewees' contact details and producing my hand-drawn figures and tables in Word or as pdfs. In the evenings, I would then write to those individuals from home to ask for interviews. I readily acknowledged her assistance, both in the thesis and in the book that I published earlier that year, as follows:

> My Personal Assistant at work, Mrs Lerato Lebeko, has helped me greatly with researching the contact details and e-mail addresses of most of my interview participants, and was kind enough to turn my hand-drawn tables and figures into decent images which accurately captured what I intended. My wife also helped with some of the tables and, more importantly, showed me how to compile them myself to the extent where I could do so unaided.

No other SEIFSA employee was ever asked – let alone "instructed" – to assist in any way during my studies.

Secondly, I did not use SEIFSA's money to buy a voice-recognition software. As far as I know, SEIFSA does not have such software. I bought such voice-recognition software for my computer at home for the purpose of transcribing interviews conducted for my PhD research. Instead, I bought, with the full knowledge and cooperation of the CFO and through him, an Olympus Digital Voice Recorder at the cost of R10 000. It is true that I used it, in addition to my cellular phone, to record my interviews for my PhD research, but it remains a SEIFSA asset and was bought for the use of the Company Secretariat to record Board and Exco Meetings. It remains in the organisation's possession to this day.

Throughout my tenure as SEIFSA CEO, I have never taken the organisation's laptop home or anywhere on business with me. Instead, I have always used my personal computer at home and my Wi-fi for work purposes, and my Samsung Notepad 10.1 when I was travelling. So, just as much as I used the digital recorder, a

SEIFSA asset, for my studies, I have also used my own assets – over a considerably longer period – for SEIFSA business.

Thirdly, as has been amply demonstrated in this book, SEIFSA was a loss-making entity when I joined it in November 2013, with the Board (then called the "Executive Committee") having approved a deficit budget of R8.1 million. We progressively reduced that deficit and, three years later, SEIFSA broke even and began generating surpluses – for three years in a row. SEIFSA was ahead of budget in the first six months of the 2019/20 financial year, but ended up registering a loss that year because of the economic shutdown occasioned by the COVID-19 pandemic. Yes, I rejected the suggestion by Trentini and Rajcoomar that we give up the west wing of the building – because it would have been unreasonable to cram SEIFSA staff members into a smaller space at a time when the COVID-19 pandemic was still raging.

Whenever possible, I have always wanted to avoid the worst form of cost-containment, staff retrenchment. I rejected a review of employee benefits because we had previously implemented such measures. As various turnaround scholars have argued, during tough times a company wants all its employees motivated and working hard, not moaning and dragging their feet because they are aggrieved. That is what yet another review of employee benefits would have accomplished. I argued that instead of having all employees aggrieved or unhappy because of a cut to their benefits (such as pension), it would be far more preferable to let a certain number of employees go and have all those who remain motivated and committed to the cause.

More importantly, there is absolutely nothing that obliged me to choose one form of cost-cutting against another. The important thing is that I had always ensured that we effect the necessary cost-containment. It mattered not that the choices made were not to the liking of Trentini, Rajcoomar, Monage, Tsehlo and Moz and that their preferred options were not adopted. All the cost-containment measures I approved meant that we were able to present a break-even budget for the 2021/22 financial year. Ultimately, that was the important thing.

Interestingly, I note in passing that, subsequent to my departure, they did, in fact, implement an even more radical round of retrenchments than I had planned. In an interview with *Engineering News* in April 2022, Trentini revealed that SEIFSA's staff complement had come down from 32 to 22, with Rajcoomar and Specht among those who had taken leave of the federation.

Fourth, it is not true that, without Board approval, I instructed Specht "to register trademarks for SEIFSA" in neighbouring countries. This was part of our three-year (2014-2017) turnaround strategy, which was approved by the Board in 2014. Bridgette Mokoetle, who was in Specht's position at the time, went ahead with efforts to register SEIFSA in Zambia, Namibia and Mozambique and payment for these registrations was made in 2015/16. Although the decision had been to register SEIFSA in additional countries each year, we put that on hold because of financial constraints at the time (we implemented our first round of retrenchments in 2015/16). However, we decided that, where efforts to register in those countries had already been made, we would continue to see through the process to the end. That is what I instructed Specht to do – and I routinely gave the Board progress reports.

Monage's desperation is at its worst when it comes to his allegation that I used SEIFSA's resources to further the interests of "the South African BRICS Manufacturing Chapter". As I have explained in this book, my involvement with the MWG of the South African Chapter of the BRICS Business Council was entirely at the instance and insistence of Monage, who had lobbied for the removal of Amka Products (Pty) Ltd Managing Director Nizam Kalla as Chairman and for him to be replaced by me. Not only did every MWG meeting or conference hosted by or at SEIFSA take place with the full knowledge of the Board of Directors, but it was also encouraged by Monage, who was himself a member of the five-person BRICS Business Council.

Fifthly, the charge concerning my alleged failure to honour SEIFSA's Values and Leadership Declaration – for both of which I am the author – is laughable. It is tantamount to the pot calling the kettle black. The fact remains that it was certain BOard

Members, as members of the SEIFSA Board, who consistently behaved in a manner inconsistent with SEIFSA's Values, as has already been explained elsewhere in this book.

Finally, I have never withheld information from the Board or failed to implement its instructions. Instead, it was Monage who, without the knowledge and approval of his fellow Board members, approved a membership discount to the CEA.

## The Disciplinary Hearing that Never Was

In addition to demanding access to a series of documents in my laptop at work, we also demanded the appointment of an independent Advocate of the High Court, to be chosen by the Chairperson of the Johannesburg Bar Council, to chair the disciplinary hearing. SEIFSA wrote back to inform us that they had appointed Advocate Van As to chair the hearing since it was their prerogative to do so. The ENS Africa team assured me that it knew Van As as "a reputable advocate" and had worked with him, hence we should have no problem with him.

The SEIFSA team dragged its feet when it came to granting us access to the documents that we needed. Closer to the hearing date, they arranged for me to be remotely connected to my work laptop, via Teamviewer, to download the documents.

When, in the preliminary meeting with both parties, Van As asked if either party would be amenable to a settlement agreement, I informed my lawyer that I was not interested in it. Instead, I wanted the hearing to proceed so that I would be vindicated at the end of the process. The hearing itself was set for 9am on 9 July 2021. I was raring for it to get started. My legal team and I had agreed on an opening statement that I would make at the beginning of the hearing, and excerpts from it follow below.

I was still unable to drive at the time because of the excruciating pain associated with the post-herpetic neuralgia, so my wife dropped me off at the offices of Fluxmans Attorneys in Rosebank. I was ushered into the room where the hearing would take place, and sitting there were Monage, Specht and Van As. When I greeted them, only Specht and Van As responded, with Monage remaining quiet. That showed me just how deeply personal the whole thing was to him.

Below are excerpts from the opening statement that I was itching to make that morning which I was denied the opportunity :

Good morning, Chairman, ladies and gentlemen

Chairman, we should not – and would not – be here this morning, were it not for a few unfortunate factors which I will touch on over the next few minutes. We are here because of the series of frivolous allegations which have been made against me. I say this because I am an honest and hard-working man of integrity who has not only served SEIFSA well over the past seven-and-a-half years, but who has also been great credit to my country, which I have served with great distinction over more than three decades.

I am a respected man who has worked for large, listed and multi-national companies more than a thousand times larger than SEIFSA, but I have never once been accused of any wrongdoing or hauled before a disciplinary hearing in any of those companies. Furthermore, I am a respected public figure who has been a Political Editor and Executive Editor of *The Star* during its heydays, Deputy Editor of *The Mercury* and Editor of *The Independent on Saturday* and the *Daily News* respectively in Durban.

When I left journalism in December 2002 after 15 years in that profession, I continued to excel at various blue-chip companies where I occupied senior, executive roles. These are companies like Anglo American, at the time the largest, hugely diversified company listed on the Johannesburg and London Stock Exchanges; American giant The Coca-Cola Company; Sasol Limited, which was then and is still now listed both here in Johannesburg and in New York; and national oil and gas company PetroSA. At all these companies, I had a reputation as an honest, hard-working and straight-talking man. My integrity has been my primary currency throughout my personal and professional lives.

Why, then, are we here this morning?

Chairman, evidence will be presented which will demonstrate that we are here because I have proved to be too honest for the liking of some key power brokers in the current Board at SEIFSA. We are here because I dared to differ with powerful men on the SEIFSA Board who have sought to use their membership of the federation's Board to get more business deals and advance their personal business interests outside SEIFSA, instead of executing their fiduciary duty to the organisation, in accordance with the dictates of Good Corporate Governance. Therefore, a frivolous case had to be made against me since I was seen as an unbending gatekeeper.

We are here because there are men with big egos and uncontrollable ambitions who are eyeing my position and who, by hook or by crook, want me to exit the organisation. We are here because these self-same men do not care about the truth, fairness or even justice. Instead, all they care about is besmirching and tarnishing my hard-earned reputation and ejecting me from SEIFSA.

Throughout my professional life, I have been a pathfinder and a pioneer and have been very mindful of my responsibility to open doors for black professionals coming after me. It does not make sense at all that I could have led such an exemplary professional life wherever I have worked, and then been so careless as to throw all that away on the twilight of my career at the smallest organisation I have ever worked for.

Chairman, we are here because, upon my arrival at SEIFSA in November 2013, I found a loss-making organisation which was a cesspool of nepotism, where my predecessors had employed their sons and some senior Executives' families were service providers to the self-same organisation. I made it very clear, right at the beginning of my tenure as CEO, that there would be no nepotism practised at SEIFSA during my tenure.

Chairman, we are here because I was uncompromising in carrying out my mandate to implement transformation at SEIFSA and yanking the organisation into the 21$^{st}$ century in terms of business practice, thus turning it around to profitability within three years.

We are here because of petty jealousy because I have successfully completed a PhD while in the employ of SEIFSA. Instead of being praised for bringing glory to the federation and being an inspiration to my staff members, I am being crucified and threatened with dismissal.

Chairman, the evidence that will be presented when we take the stand will show that there are certain individuals within the SEIFSA Board who are the primary conspirators and who have prevailed upon some staff members to put together the various allegations of misconduct on my part so that I can be removed from my position as CEO since I am an incorrigible gatekeeper. We will prove this when I give evidence in support of my case and when former Board Chairman Alpheus Ngapo, who has known me the longest and was on that Board when I joined the federation in November 2013, takes the stand as a witness. I will point out many more reasons why these co-conspirators want me out of SEIFSA.

In conclusion, Chairman, I am innocent and plead not guilty to the hotchpotch of frivolous charges laid against me.

Sadly, I did not get to deliver this opening statement. I did not do so because, yet again, Van As placed the possibility of a settlement agreement on the table – and SEIFSA jumped at the opportunity. They said that they would withdraw the charges, provided that we could reach a settlement. During a side-consultation, my lawyer advised me to consider what SEIFSA was prepared to place on the table for a parting of ways. He warned that it was clear that the relationship between the Monage-led Board and I had broken down irretrievably and informed me that, if the matter went all the way to the Labour Court, I might find myself awarded no more than six months' remuneration and benefits as a separation package. Therefore, if we could reach such an agreement now, that would be a victory, and he would advise me to settle.

Within an hour, an agreement was reached: SEIFSA would withdraw its fabricated charges, issue a public statement to that effect and pay me to walk away. After to-ing and fro-ing, an agreement much better than the one that the ENS Africa team had indicated as a possibility was reached.

Petty to the end, Monage turned down our suggestion that I would return to the office for a few days and then tender my resignation. Instead, he insisted on me agreeing to resign immediately, with SEIFSA to release a statement vindicating me and announcing my resignation. That statement, as issued on 23 July 2021, is reproduced below.

## SEIFSA

Steel and Engineering Industries Federation of Southern Africa
OUR PASSION, YOUR SUCCESS

**SEIFSA CEO KAIZER NYATSUMBA RESIGNS**

JOHANNESBURG, 23 JULY 2021 – SEIFSA confirms that it has uplifted the suspension of its Chief Executive Officer Kaizer Nyatsumba, who was placed on precautionary suspension at the beginning of May, SEIFSA Board Chairman Elias Monage announced yesterday.

Due to recent events, the parties reached an agreement to mutually separate. Dr Nyatsumba has decided to return to his consulting company, KMN Consulting, following completion of his PhD.

SEIFSA wishes Dr Nyatsumba well in his future endeavours.

**Ends**

**Issued by:**

Nuraan Alli

Acting Marketing, Sales and Communications Executive

In a separate statement sent to News24, to which news of my suspension was leaked, I added the following paragraph:

I have enjoyed my time with SEIFSA, where I was the first-ever black (African) CEO in the federation's history. I am glad that, during my tenure, we were able to stem SEIFSA's losses, following the expiry of the Collective Bargaining Levy in 2012, and to turn the federation around to produce surpluses three years in a row, until the advent of the COVID-19 pandemic last year.

While Monage captured the sequence of events correctly in his Chairman's Report in SEIFSA's Annual Report of that year, Company Secretary Louwresse Specht (2021:75) did not. Instead, she created the erroneous impression that my resignation was occasioned by the act of my precautionary suspension and failed to reflect that I resigned only after the suspension was lifted. "During May 2021," she wrote, "the CEO of SEIFSA, Dr Kaizer Nyatsumba, was placed on precautionary suspension and in July 2021 he resigned from SEIFSA". It is very strange that an admitted attorney, who was intricately involved in the process on the side of SEIFSA, would make such a basic mistake, without her CEO spotting and correcting it.

Once the news of my parting of ways with SEIFSA broke, I received the following email from Ross Williams, sent to both my private email addresses:

## (no subject)

**RW**  Ross Williams <ross@successioncapital.ie>
To  Kaizer Nyatsumba; Kaizer Nyatsumba

So delighted to hear you have finally been disposed of!!
And by Elias of all people.
Truly hilarious!

Sent from my iPhone

*Source: Email sent to the author by Ross Williams*

Indeed, through guile, Monage – the third black President in SEIFSA's history – had delivered the *coup de grace*. He had got rid of the federation's first – and so far only – black CEO, thus succeeding, at a considerable cost to SEIFSA, where some of his predecessors had failed spectacularly.

## Conclusion

The Steel and Engineering Industries Federation of Southern Africa (SEIFSA) turned 80 in 2023. I have been privileged to have been part of its most recent decade, and to have played a role in its transformation and its financial rehabilitation following the expiry of the CBL in December 2012.

Sadly, neither the M&E sub-sector nor the larger manufacturing sector to which it belongs has recovered fully since the global financial crisis of 2008/9. Indeed, the irony remains that the manufacturing sector thrived during the apartheid era, when punitive economic sanctions were still in place against an isolated South Africa that had erected high import-tariff barriers and has struggled to deal with international competition during the democratic era. With manufacturing generally acknowledged as a major creator of jobs, it remains to be seen whether the Government's efforts to stimulate the sector – perhaps through the mooted infrastructure projects – will have the desired effect of lowering the country's hight unemployment levels.

I am grateful to have had the opportunity to be part of SEIFSA and am proud to have played a role in its transformation and financial turnaround. I wish the organisation well for the future, and I hope that in the years to come it will have leaders whose primary loyalty is to its members and related stakeholders – and not to themselves.

# References

Ade, M. & Kruger, M. (2019). *State of the Metals and Engineering Sector Report 2018-2019*. Johannesburg: Steel and Engineering Industries Federation of Southern Africa.

America's Investment in the Future. Available from: https://www.nsf.gov/about/history/nsf0050/pdf/manufacturing.pdf

Angus, B. (2022). Letter: Who's the boss in the steel sector? *Financial Mail*. Available from: https://www.businesslive.co.za/fm/opinion/letters/2022-11-03-letter-whos-the-boss-in-the-steel-sector/

BakerBaynes.com. (2017). The History and Future of Manufacturing in South Africa. Available from: https://www.bakerbaynes.com/history-future-manufacturing-south-africa/

Barnes, J., Black, A. & Techakanont, K. (2017). Industrial Policy, Multinational Strategy and Domestic Capability: A Comparative Analysis of the Development of South Africa's and Thailand's Automotive Industries. *The European Journal of Development Research*, 29(1), 37-53. https://doi.org/10.1057/ejdr.2015.63

Bhorat, H. & Rooney, C. (2017). State of Manufacturing in South Africa. Working Paper 201702. A DPRU Working Paper commissioned for MerSeta. 201702. Development Policy Research Unit. ISBN 978-1-920633-41-7

Boppart, T. (2014). Structural Change and the Kaldor Facts in a Growth Model with Relative Price Effects and Non-Gorman Preferences. *Econometrica*, 82(6), 2167–2196. https://doi.org/10.3982/ECTA11354

Callinicos, L. (1987). *Working Life –1886-1940*. Johannesburg: Raven Press (Pty) Ltd.

CFI. (2022). Manufacturing. Available from: https://corporatefinanceinstitute.com/resources/valuation/manufacturing/

Carson, D. (2011). Report of the Executive Director 2011. Johannesburg: SEIFSA

Chamber of Mines South Africa. (2013). The South African mining sector in 2013. Available from: https://www.mineralscouncil.org.za/industry-news/publications/facts-and-figures

Chamber of Mines South Africa. (2017). Facts and Figures 2016. Available from: https://www.mineralscouncil.org.za/industry-news/publications/facts-and-figures

Chalmers, A. (2016). The demise of the MEIBC: It only has itself to blame. Available from: https://www.linkedin.com/pulse/demise-metal-industry-bargaining-council-meibc-only-has-scheepers

Chibanguza, T. (2013). *State of the Metals and Engineering (M&E) Sector 2023.* Johannesburg: Steel and Engineering Industries Federation of Southern Africa.

Coffin, D.A. (2003). The State of Steel. *Indiana Business Review*, Spring.

*Concise Oxford English Dictionary* (10th ed., revised). Oxford: Oxford University Press.

Coutsoukis, P. (2004). South African Manufacturing. Available from: https://photius.com/countries/south_africa/economy/south_africa_economy_manufacturing.html

Corbett, M. (n.d.). Oil Shock of 1973-74. Available from: https://www.federalreservehistory.org/essays/oil-shock-of-1973-74#:~:text=October%201973%E2%80%93January%201974&text=The%20embargo%20ceased%20U.S.%20oil,a%20barrel%20in%20January%201974.

Corporate Finance Institute. (2022). Manufacturing. Available from: https://corporatefinanceinstitute.com/resources/valuation/manufacturing/

Dasgupta, S. & Singh, A. (2007). "Manufacturing, Services and Premature Deindustrialization in Developing Countries: A Kaldorian Analysis" In Mavrotas. G. & Shorrocks, A. (Eds.). *Advancing Development* (pp. 435–454). Palgrave Macmillan UK. https://doi.org/10.1057/9780230801462_23

Department of State. United States of America Office of the Historian. Oil embargo 1973-74. Available from: https://history.state.gov/milestones/1969-1976/oil-embargo

De Vries, G., Arfelt, L., Drees, D., Godemann, M., Hamilton, C., Jessen-Thiesen, B., Kaya, A. I., Kruse, H., Mensah, E., & Woltjer, P. (2021). The Economic Transformation Database (ETD): Content, sources, and methods. UNU-WIDER. https://doi.org/10.35188/UNUWIDER/WTN/2021-2

Dondofema, R.A., Matope, S. & Akdogan.G. (2017). South African Iron and Steel Industrial Evolution: An Industrial Engineering Perspective. *South African Journal of Industrial Engineering*, 28(4), pp. 1-13. https://doi.org/10.7166/28-4-1683

Du Plessis, G. (2024). MEIBC's deceptive wage offer bad news for skilled employees – Solidarity. Available from: https://www.politicsweb.co.za/politics/meibcs-deceptive-wage-offer-bad-news-for-skilled-e

Encyclopedia, Science News & Research Reviews. Steel Crisis. Available from: https://academic-accelerator.com/encyclopedia/steel-crisis

*Engineering News*. Business Leader: Lucio Trentini. Available from: https://www.engineeringnews.co.za/article/lucio-trentini-2022-04-01

Eyewitness News. (2014). Zuma's large Cabinet "a waste of money". Available from: https://ewn.co.za/2014/05/26/Zumas-large-Cabinet-will-waste-money

Fine, B. & Rustomjee, Z. (2018). *The Political Economy of South Africa: From Minerals–Energy Complex to Industrialisation* (1st ed.). Routledge. https://doi.org/10.4324/9780429496004

Fortunato, A. (2022). Getting Back on the Curve: South Africa's Manufacturing Challenge. The Growth Lab at Harvard University. https://doi.org/10.2139/ssrn.4411854

Gumede, M. (2023). SEIFSA warns of bleak year ahead for metals and engineering. *Business Day*, p3, 17 February 2023.

Hausmann, R., & Rodrik, D. (2002). Economic Development as Self-Discovery (No. w8952; p. w8952). National Bureau of Economic Research. https://doi.org/10f.3386/w8952

IndustriALL. (2023). Union wins as strike action ends in three-year wage agreement. Available from: https://www.industriall-union.org/union-wins-as-strike-action-ends-in-three-year-wage-agreement

International Trade Administration Commission of South Africa. (n.d.). An Overview of ITAC. Available from: http://www.itac.org.za/pages/about-itac/an-overview-of

International Trade Administration. (2023). South Africa – Country Commercial Guide. Available from: https://www.trade.gov/country-commercial-guides/south-africa-import-tariffs

Jacobs, A. (1948). *South African Heritage: A Biography of H.J. van der Bijl.* Pietermaritzburg: Shuter and Shooter.

Kaplan, D. (2004). Manufacturing in South Africa over the last decade: a review of industrial performance and policy. *Development Southern Africa*, 21(4), 623-544. https://doi.org/10.1080/0376835042000288824

Kenward, L.R. (1987). The Decline of the US Steel Industry. Why Competitiveness Fell Against Foreign Steel Makers. *Finance & Development.* Available from: https://www.elibrary.imf.org/view/journals/022/0024/004/article-A009-en.xml

Kilian, N. (2016). Solidarity urges Oliphant to intervene at MEIBC. Available from: https://www.engineeringnews.co.za/print-version/solidarity-urges-oliphant-to-intervene-at-meibc-2016-05-06

MacKay, D. (2023). SA steel industry and state joined in embrace that has lasted a century. *Business Day*, 7 February 2023.

Macrotrends. (n.d.) South Africa Manufacturing Output 1960-2022. Available from: https://www.macrotrends.net/countries/ZAF/south-africa/manufacturing-output

Mhlanga, I. (2023). Manufacturing sector gets to grips with load-shedding. *Business Day*, p. 7, 7 July 2023.

MEIBC. (2011). Constitution. Available from: https://www.meibc.co.za/images/PDFS/MEIBC_CONSTITUTION_CURRENT.pdf

Monage, E. (2023). The Collapse of SA's Steel Master Plan and disintegration of industry. [Online}. *Engineering News.* Available from: https://www.engineeringnews.co.za/article/opinionthe-collapse-of-the-steel-master-plan-and-disintegration-of-industry-2023-12-01

Mongalo, B. (2023). Investment in manufacturing will grow pensions, jobs. *Sunday Times Business Times*, p.7, 12 June 2023.

MyBroadband. 2015. South African Electricity Prices – 1994 to 2015. [Online]. Available: http://mybroadband.co.za/news/energy/130320-south-african-electricity-prices1994-to-2015.htm.

National Science Foundation. [n.d.] *Manufacturing: the form of things unknown*. Available from: https://www.nsf.gov-about-history.nsf0050

National Planning Commission. 2011. National Development Plan. Pretoria: Government Printer.

Papenfus, G. (2016). The Demise of the MEIBC: It only has itself to blame. Available from: https://www.bizcommunity.com/Article/196/646/143569.html

Photius.com. (1996). South Africa Manufacturing. Available from: https://photius.com/countries/south_africa/economy/south_africa_economy_manufacturing.html

Plunkert, L. M. (1990). The 1980's [sic]: a decade of job growth and industry shifts. *Monthly Labour Review*, September 1990.

Ryan, C. (2022). Nearly R200bn strategic infrastructure projects are under construction. *Moneyweb*. Available from: Available from: https://www.moneyweb.co.za/news/industry/nearly-r200bn-strategic-infrastructure-projects-are-under-construction/?bid=120_10977_8149987

Reuters. (2014). Ghana bans export of ferrous scrap metal to support local industry. Available from: https://www.reuters.com/article/idUSL6N0N7343/

Rodrik, D. (2006). Understanding South Africa's Economic Puzzles. Bureau for Research and Economic Analysis of Development (BREAD), Working Paper No. 131. https://doi.org/10.3386/w12565

Rodrik, D. (2015). Premature Deindustrialization (No. w20935; p. w20935). National Bureau of Economic Research. https://doi.org/10.3386/w20935

Rowe, D. (2016). Lessons from the steel crisis of the 1980s. *The Conversation*. Available from: https://theconversation.com/lessons-from-the-steel-crisis-of-the-1980s-57751

Sablik, T. (n.d.). Recession of 1981-82. Available from: https://www.
federalreservehistory.org/essays/recession-of-1981-
82#:~:text=Both%20the%201980%20and%201981,known%20
as%20the%20Phillips%20Curve.

Schwab, K. (2016). The Fourth Industrial Revolution: what it means,
how to respond. Available from: https://www.weforum.org/
agenda/2016/01/the-fourth-industrial-revolution-what-it-
means-and-how-to-respond/

SEIFSA. (n.d.). History of SEIFSA. [Online]. Available from www.seifsa.
co.za.

SEIFSA Annual Review. (2013). Celebrating 70 Years of Service Excellence.
Johannesburg: SEIFSA. Available from: https://seifsa.co.za/wp-
content/uploads/2017/06/2013Oct02_SEIFSA_Annual_Review_
Small.pdf

SEIFSA. (2015). Expiry of the MEIBC Administration and Expenses
Levy and Dispute Levy Agreement 31 March 2015. Available
from: https://www.seifsa.co.za/article/expiry-of-the-meibc-
administration-and-expenses-levy-agreement-and-dispute-
levy-agreement-31-march-2015/

SEIFSA. (2017). Letter to Stakeholders: Solidarity Judgement. Available
from: https://www.seifsa.co.za/article/letter-to-stakeholders-
solidarity-judgement/

Seth, S. (2022). How China Impacts the Global Steel Industry.
Investopedia. Available from: https://www.investopedia.com/
articles/investing/021716/how-china-impacts-global-steel-
industry.asp

Specht, L. (2021). Company Secretariat Report. SEIFSA Annual
Review 2021. Available from: https://issuu.com/
coloursvary/docs/2021september-17_seifsa-annual-
review_1?fr=sNTRkNjUyOTY0Mzg

South African History Online. (n.d.). 2014 South African platinum strike:
Longest wage strike in South Africa. Available from: https://
www.sahistory.org.za/article/2014-south-african-platinum-
strike-longest-wage-strike-south-africa

South African Press Association. (2014). Court sets aside wage agreement – NEASA. Available from: https://www.engineeringnews. co.za/print-version/court-sets-aside-wage-agreement---neasa-2014-12-18

South African Reserve Bank (SARB). (1993). Annual Economic Report – 1993. Available from: https://www.resbank.co.za/en/home/ publications/publication-detail-pages/reports/annual-economic-reports/1993/5550

South African Reserve Bank (SARB). (2016). Economic and Financial Data for South Africa. Available from: http://wwwrs.resbank.co.za/ webindicators/EconFinDataForSA.aspx#RealSector.

Swanepoel, C. & Fliers, P.T. (2021). The fuel of unparalleled recovery: Monetary policy in South Africa between 1925 and 1936. *Economic History of Developing Regions*, 36(2). https://doi.org/10.1080/2078 0389.2021.1945436

Tarr, D.G. (1988). The Steel Crisis in the United States and the European Community: Causes and Adjustments. In Baldwin, Robert. E., Hamilton, Carl. B. & Sapir, A. (Eds). *Issues in US-EC Trade Relations.* Illinois: University of Chicago Press.

Tregenna, F. (2008). The Contributions of Manufacturing and Services to Employment Creation and Growth in South Africa. *South African Journal of Economics*, 76(2). https://doi.org/10.1111/j.1813-6982.2008.00187.x

Trentini, L. (2024). SEIFSA and NUMSA conclude Historic Agreement in record time. *2024 Wage Negotiations Update.* Available from: https://www.seifsa.co.za/article/category/wage-negotiations-updates/

US Department of State. Oil Embargo, 1973-1974. (n.d.) Available from: https://history.state.gov/milestones/1969-1976/oil-embargo

United States National Science Foundation. Manufacturing – the form of things unknown. [Online]. Available from: https://atecentral.net/ r21501/manufacturing_the_form_of_things_unknown

Van Biljon, D. (2014). SEIFSA: 70 YEARS ON – Passion for SA's development still a major driver. *Financial Mail*, 28 March 2014.

Whitney, J.O. (1987). Turnaround Management Every Day. *Harvard Business Review*, September Issue. Available from: https://hbr.org/1987/09/ turnaround-management-every-day

Williams, G., Cunningham, S. & De Beer, D. (2014). Advanced Manufacturing and Jobs in South Africa: An Examination of Perceptions and Trends. Paper presented at the International Conference on Manufacturing-Led Growth for Employment and Equality, 20-21 May.

Wilson, J.M. (2000). History of Manufacturing Management. In Swamidass, P.M. (eds). *Encyclopedia of Production and Manufacturing Management.* Bostron: Springer. Available from: https://doi.org/10.1007/1-4020-0612-8_406

Zingoni, T. (2020). Deconstructing South Africa's construction industry performance. *Mail & Guardian*, 19 October 2020. Available from: https://mg.co.za/thought-leader/opinion/2020-10-19-deconstructing-south-africas-construction-industry-performance/

Zwane, T. (2023). SA's factory output in decline for 16 years. *Business Day*, p.1, 25 July 2023.

# Other books by
# Kaizer Mabhilidi Nyatsumba

*On The Precipice* (A Novel)

*Successfully Implementing Turnaround Strategies in State-Owned Companies: SAA, Kenya Airways and Ethiopian Airlines as Case Studies*

*South Africa as an International Brand: An Assessment of How it can be Marketed More Effectively*

*Incomplete Without My Brother, Adonis*

*ALL SIDES OF THE STORY: A Grandstand View of South Africa's Political Transition*

*Nelson Mandela: The South African Leader who was Imprisoned for Twenty-Seven Years for Fighting Against Apartheid* (2nd edition, co-written with Benjamin Pogrund)

*Desmond Tutu: The Brave and Eloquent Archbishop Struggling Against Apartheid in South Africa* (3rd edition, co-written with David Winner)

*UMLOZI* (Zulu poetry)

*The Shaman* (Poetry)

*When Darkness Falls* (Poetry)

*Silhouettes* (Poetry)

*In Love with a Stranger* (Short stories)

*A Vision of Paradise* (Short stories)

www.ingramcontent.com/pod-product-compliance
Lightning Source LLC
Chambersburg PA
CBHW061156240326
R18026500001B/R180265PG41519CBX00011B/15

* 9 7 8 0 9 0 6 7 8 5 3 7 9 *